Professionalism
Skills for Workplace Success

Fourth Edition

Lydia E. Anderson

Sandra B. Bolt

PEARSON

Boston   Columbus   Hoboken   Indianapolis   New York   San Francisco
Amsterdam   Cape Town   Dubai   London   Madrid   Milan   Munich   Paris   Montréal   Toronto
Delhi   Mexico City   São Paulo   Sydney   Hong Kong   Seoul   Singapore   Taipei   Tokyo

Senior Acquisitions Editor: Paul Smith
Lead Program Manager: Lauren Finn
Development Editor: Claire Hunter
Executive Marketing Manager: Amy Judd
Executive Digital Producer: Stefanie Snajder
Digital Editor: Tracy Cunningham
Media Producer: Kate Goforth
Content Specialist: Celeste Kmiotek
Project Manager: Shannon Kobran

Project Coordination and Text Design: MPS North America, LLC
Page Makeup: Laserwords
Design Lead: Heather Scott
Cover Designer: Studio Montage
Cover Image: Cherezoff / Shutterstock
Senior Manufacturing Buyer: Roy L. Pickering, Jr.
Printer/Binder: Courier/Kendallville
Cover Printer: Courier/Kendallville

Acknowledgments of third-party content appear on page 287, which constitute an extension of this copyright page.

PEARSON, ALWAYS LEARNING, and MYSTUDENTSUCCESSLAB are exclusive trademarks, in the United States and/or other countries, owned by Pearson Education, Inc. or its affiliates.

Unless otherwise indicated herein, any third-party trademarks that may appear in this work are the property of their respective owners and any references to third-party trademarks, logos, or other trade dress are for demonstrative or descriptive purposes only. Such references are not intended to imply any sponsorship, endorsement, authorization, or promotion of Pearson's products by the owners of such marks, or any relationship between the owner and Pearson Education, Inc., or its affiliates, authors, licensees, or distributors.

10 9 8 7 6 5 4 3 2 1—CRK—17 16 15 14

www.pearsonhighered.com

Student Edition ISBN 10: 0-32-195944-2
Student Edition ISBN 13: 978- 0-32-195944-7
A la Carte ISBN 10: 0-13-386894-X
A la Carte ISBN 13: 978-0-13-386894-4

Brief Contents

Lydia E. Anderson has a master's degree in business administration with an emphasis in marketing. In addition to years of corporate marketing and strategic planning experience, she has been teaching for over eighteen years in both community college and university settings. She is currently a tenured faculty member and former chair of the Business Administration and Marketing Department at Fresno City College in Fresno, California. She also serves as an adjunct professor at California State University, Fresno. Her teaching areas of expertise include human relations in business, management, supervision, human resource management, and marketing. Ms. Anderson is also active in (California) statewide business curriculum development, student success and enrollment management initiatives, and in Academic Senate. She regularly consults with corporations on business topics relating to management and marketing to ensure currency in instruction.

Sandra B. Bolt has a master's degree in business administration with an emphasis in human resource management. She has been teaching in the college setting for over twenty-four years. She is currently a tenured faculty member and past chair of the Business and Technology Department at Fresno City College in Fresno, California. Her teaching areas of expertise include workplace relationships, office occupations, office technology, résumé/interview, business communication, document formatting, and computer applications. She currently serves as the Secretary-Treasurer of the State Center Federation of Teachers. She has extensive secretarial, treasurer, and leadership experience and has served as a computer applications trainer. She has led personal financial management sessions for community groups. She has been a volunteer guest speaker at professional conferences and high school career fairs, in addition to her involvement with committees and student functions at Fresno City College.

Both authors have used their professional, educational, and personal experiences to provide readers with realistic stories and challenges experienced in a typical workplace.

Dedication

To all those looking for the job of your dreams—may God richly bless you as you reach your professional and personal goals and ultimately discover that success starts from within.

—Lydia E. Anderson

I dedicate this fourth edition to everyone who has used this text for improving professionalism in the workplace. I thank my husband, Bret, for being supportive, loving, and my best friend; and to my son, Brandon, who has made me proud by growing into such a wonderful young man. In addition, I dedicate this text to the memory of my parents and to God, for giving me the blessings and strength I have in my life.

—Sandra B. Bolt

Contents

Module 2 Workplace Basics

Module 3   Relationships

Module 4   Career Planning Tools

Acknowledgments

The success of our first three editions exceeded our wildest dreams, and we are tremendously thankful to those who have included our materials in their educational toolboxes. This fourth edition continues to integrate feedback from business leaders and educators who openly share their expertise, ideas, and concerns regarding necessary workplace skills and expected behaviors specific to today's tumultuous economic environment. This fourth edition continues to prepare students for real-world success and contributes to employer and economic success as well. We remain committed to providing readers a competitive advantage in successfully realizing and achieving their career goals and believe this updated edition does just that.

Our continued thanks to Pearson Education for providing us this opportunity. We are specifically grateful to Claire Hunter for her consistency and support.

Our grateful appreciation goes to the reviewers of this text for their honest and valued insight:

Dennis Nasco, Jr., Southern Illinois University
Susan Moak Nealy, Baton Rouge Community College
Heidi Lee Arrington, University of Hawai'i - Kapi'olai Community College
Brian Penberty, ECPI College of Technology
Laura Portolese Dias, Shoreline Community College
Dr. Raven Davenport, Houston Community College
Denver L. Riffe, National College
Thomas R. Smith, Prism Career Institute
Zachary Stahmer, Anthem Education Group
Anita Wofford, The Art Institute of Charlotte
Michael Bailey, Texas A&M University
Kevin Michael Bratton, West Georgia Technical College
Laure S. Burke, Kapiolani Community College
Kelley Ashby, The University of Iowa
Martha Hunt, NHTI-Concord's Community College
Denise Doyle, Institute of Technology
Regina D. Hartley, Ph.D., Caldwell Community College & Technical Institute
Tammy Mohler-Avery, Brandford Hall Career Institute
Gina P. Robinson, Brunswick Community College

Preface

Professionalism Skills for Workplace Success, fourth edition, continues to remain committed to its original purpose of addressing employer concerns by providing those new to the workplace with basic skills for success on the job and providing job seekers the tools they need to secure the job of their dreams. This unique text prepares students for their first professional workplace experience by linking an individual's life plan to behavior necessary for workplace success. The text content is applicable to any individual who will be transitioning from the classroom to the work environment. The book is designed not just as a textbook, but as a workbook to be kept and referred to throughout one's career.

While economics and technology continued to be the primary influence for revisions in this fourth edition, the authors also strived for a broader reach by addressing contemporary workplace issues and providing examples for careers that occur inside and outside a traditional office setting. Our world continues to struggle with challenging economics and historic unemployment rates. Therefore, it is imperative that job seekers and those new to the workplace not only demonstrate, but exceed expectations regarding business etiquette, appropriate technology use, and proper workplace attire. Students also need to understand how these expectations differ from personal social situations. In today's increasingly competitive work environment, it is essential that students communicate in a professional manner, maturely deal with conflict, and be accountable team members, consistently behaving in a fair and ethical manner. This fourth-edition text continues to address these issues by providing readers with realistic, current, and practical skills necessary to maintain success on the job.

Unique Approach

Professionalism Skills for Workplace Success, fourth edition, extends beyond a typical student success or résumé/job search text. The foundation of this text assists students in creating a life plan that addresses short- and long-term personal, professional/career, and financial goals. The text then provides students practical skills and challenges to immediately begin implementing behaviors that support their life plan, with a primary emphasis on professional/career behaviors. The end of each chapter provides activities that promote student success and relates chapter content back to a student's life plan. Text content seamlessly emphasizes the relationship between job search/résumé development and human relations in the workplace. This fourth edition continues to integrate input from industry leaders, and addresses timely and critical topics students need to know when transitioning from campus to career. Presented in a simple, highly interactive format, this fourth-edition text assists individuals in understanding the foundation of effective workplace relationships and how to appropriately manage these relationships toward career success. Beginning with the basic management principle that individual workplace performance affects organizational success and profitability readers are able to integrate soft skills within the framework of a formal business structure. The topics and principles presented benefit individuals in any industry and career. Utilizing Topic Situations embedded in each chapter, students gain valuable insights from real workplace dilemmas presented as mini–case studies. Each Topic Situation is followed by a Topic Response that poses questions on how best to handle these challenging dilemmas. Back by popular demand are introductory assessments, in-chapter exercises, talk it outs (student discussion topics), and valuable end-of-chapter activities designed to improve the reader's understanding and application of the material through written and oral communication skill building assignments.

This book is written for individuals wanting to excel in their career. Attitude, communication, and human relations are the keys to surviving in today's challenging, competitive, and uncertain workplace. The text teaches realistic career building skills and motivates individuals toward improving both personal and professional performance.

Organization

When used in its entirety, the text is divided into four modules and sixteen chapters that are arranged to accommodate quarter-, half-, or full-semester courses taught in a traditional face-to-face classroom, online, or in a hybrid structure. Modules include:

Module 1: Self-Management
Module 2: Workplace Basics
Module 3: Relationships
Module 4: Career Planning Tools

When using selected chapters, it is strongly recommended that complimentary chapters accompany each other to provide complete content coverage. Complimentary chapters include:

- "Communication" (chapter 9) and "Electronic Communications" (chapter 10)
- "Motivation, Leadership, and Teams" (chapter 11) and "Conflict and Negotiation" (chapter 12)
- "Job Search Skills" (chapter 13), "Résumé Package" (chapter 14), and "Interview Techniques" (chapter 15)

New to This Edition

The text blends career goals and workplace relations throughout, emphasizing three pillars of teaching and learning: life planning, workplace skills, and career planning.

Life Planning offers enhanced learning outcomes, in-chapter exercises, and end-of-chapter activities to support Bloom's Taxonomy and help students think critically about their life and career goals.

- **Summary words and revised learning outcomes** provide introduction to topic and content focus.
- **In-chapter exercises have been updated throughout including changing** *Cory stories* to *Topic Situation* and *Topic Response* case studies, *Web Searches*, "*Think About It*" **and** "*Talk About It*" **discussion and**

reflection questions promote discussion, provide a means for topic clarification, and immediate application of content.

- **Enhanced end-of-chapter** *Concept Review and Application section* reinforces student learning by providing hands-on application of topic content. These include *Workplace Do's and Don'ts, Summary of Key Concepts, Key Terms* (set up as a Self-Quiz to be matched with definitions), *Think Like a Boss,* and new/revised *End of Chapter Activities*

- **How-Do-You-Rate?** Assessments provide students an introduction to the chapter topic through brief, fun, and realistic applications.

Workplace Skills provide new and enhanced content to address employer concerns related to millennial and reentry students.

- **Personal branding content and increased focus on quality and accountability** help students recognize the importance of immediately identifying a personal brand and integrating professional behaviors that represent that brand.

- **Discussion on student loans, cash management, and online protection on identity theft** addresses personal finance issues today's students face

- **Increased discussion, examples, and case studies on communication/technology use etiquette** addresses employer concern that today's students lack basic communication, spelling, and grammatical skills and are too reliant on communication devices.

Career Planning features revised career planning to address current market conditions.

- **Key topics have been reordered, including completion of accomplishments worksheet, writing of career objective/personal profile, and personal commercial** streamlines the process and better identifies key skill sets.

- **New and enhanced checklists for job search tools and processes** ensure students include/address key elements necessary for the job search portfolio, interview portfolio, interview preparation, and post-interview activities.

- **Advanced Skill Set Résumé Format** provides a résumé format for those with extensive career experience. This format highlights, communicates, and sells specific job skills and work accomplishments, and replaces the chronological format.

Also Available with MyStudentSuccessLab™

This title is also available with MyStudentSuccessLab—an online homework, tutorial, and assessment program designed to work with this text to engage students and improve results. Within its structured environment, students practice what they learn, test their understanding, and pursue a plan that helps them better absorb course material and understand difficult concepts.

Personalize Learning with MyStudentSuccessLab

This learning outcomes-based technology promotes student engagement through:

- **Full Course Pre- and Post-Diagnostic** test based on Bloom's Taxonomy linked to key learning objectives in each topic.
- Each individual topic in the Learning Path offers a **Pre- and Post-Test** dedicated to that topic, an **Overview** of objectives to build vocabulary and repetition, access to **Video interviews** to learn about key issues 'by students, for students', **Practice** exercises to improve class prep and learning, and **Graded Activities** to build critical thinking skills and develop problem-solving abilities.
- **Student Resources** include Finish Strong 247 YouTube videos, Calculators, and Professionalism/Research & Writing/Student Success tools.
- **Student Inventories** increase self-awareness, and include *Golden Personality* (similar to Meyers Briggs, gives insights on personal style), *Conley Readiness Index (CRI)* (measures readiness and likelihood for success, gives insight into student aspirations).
- **Title-specific version available** as an option for those who teach closely to their text. This course would include the national eText, Chapter specific quizzing, and Learning Path modules that align with the chapter naming conventions of the book.

Instructor Resources

Online Instructor's Manual

(www.pearsonhighered.com/irc)

This manual provides a framework of ideas and suggestions for activities, journal writing, thought-provoking situations, and online implementation including MyStudentSuccessLab recommendations.

Online PowerPoint Presentation

(www.pearsonhighered.com/irc)

A comprehensive set of PowerPoint slides that can be used by instructors for class presentations and also by students for lecture preview or review. The Power-Point presentation includes summary slides with overview information for each chapter to help students understand and review concepts within each chapter.

MyStudentSuccessLab

(www.mystudentsuccesslab.com)

This title is also available with MyStudentSuccessLab—an online homework, tutorial, and assessment program designed to work with this text to engage students and improve results. Within its structured environment, students practice what they learn, test their understanding, and pursue a plan that helps them better absorb course material and understand difficult concepts. Beyond the Full Course Pre- and Post-Diagnostic assessments, and Pre- and Post-tests within each module, additional learning outcomes-based tests can be created/selected using a secure testing engine, and may be printed or delivered online.

If interested in adopting this title with MyStudentSuccessLab, ask your Pearson representative for the correct package ISBN and course to download.

Course Redesign

(www.pearsoncourseredesign.com)

You deliver instruction, measure the results of your course redesign, and get support for data collection and interpretation.

Implementation and Training

(www.mystudentsuccesscommunity.com)

Access MyStudentSuccessLab training resources such as Best Practices implementation guide, How Do I videos, Self-paced training modules, and 1:1 Expert on Demand sessions with a Faculty Advisor, and videos, posts, and communication from student success peers.

CourseConnect

(www.pearsonlearningsolutions.com/courseconnect)

This title is also available with CourseConnect-designed by subject matter experts and credentialed instructional designers; it offers customizable online courses with a consistent learning path, available in a variety of learning management systems as self-paced study.

CourseSmart Textbooks Online

(www.coursesmart.com)

As an alternative to purchasing the print textbook, students can subscribe to the same content online and save up to 50% off the suggested list price of the print text. With a CourseSmart e-textbook, students can search the text, make notes online, print out reading assignments that incorporate lecture notes, and bookmark important passages for review.

Custom Services

(www.pearsonlearningsolutions.com)

With this title, we offer flexible and creative choices for course materials that will maximize learning and student engagement. Options include Custom library, publications, technology solutions, and online education.

Professional Development for Instructors

(www.pearsonhighered.com/studentsuccess)

Augment your teaching with engaging resources. Visit our online catalog for our Ownership series, Engaging Activities series, and Audience booklets.

Resources for Your Students

(www.pearsonhighered.com/studentsuccess)

Help students save and succeed throughout their college experience. Visit our online catalog for options such as Books a la Carte, CourseSmart eTextbooks, Pearson Students program, IDentity Series, Success Tips, and more.

Pearson Course Redesign

Collect, measure, and interpret data to support efficacy.

Rethink the way you deliver instruction.

Pearson has successfully partnered with colleges and universities engaged in course redesign for over 10 years through workshops, Faculty Advisor programs, and online conferences. Here's how to get started!

- Visit our course redesign site at www.pearsoncourseredesign.com for information on getting started, a list of Pearson-sponsored course redesign events, and recordings of past course redesign events.

- Request to connect with a Faculty Advisor, a fellow instructor who is an expert in course redesign, by visiting www.mystudentsuccesslab.com/community.

- Join our Course Redesign Community at www.community.pearson.com/courseredesign and connect with colleagues around the country who are participating in course redesign projects.

Don't forget to measure the results of your course redesign!

Examples of data you may want to collect include:

- Improvement of homework grades, test averages, and pass rates over past semesters

- Correlation between time spent in an online product and final average in the course

- Success rate in the next level of the course

- Retention rate (i.e., percentage of students who drop, fail, or withdraw)

Need support for data collection and interpretation?

Ask your local Pearson representative how to connect with a member of Pearson's Efficacy Team.

MyStudentSuccessLab
Help students start strong and finish stronger.

MyStudentSuccessLab™

MyLab from Pearson has been designed and refined with a single purpose in mind—to help educators break through to improving results for their students.

MyStudentSuccessLab™ (MSSL) is a learning outcomes-based technology that advances students' knowledge and builds critical skills, offering ongoing personal and professional development through peer-led video interviews, interactive practice exercises, and activities that focus on academic, life, and professional preparation.

The **Conley Readiness Index (CRI), developed by Dr. David Conley, is now embedded in MyStudentSuccessLab.** This research-based, self-diagnostic online tool measures college and career readiness; it is personalized, research-based, and provides actionable data. Dr. David Conley is a nationally recognized leader in research, policy, and solution development with a sincere passion for improving college and career readiness.

Developed exclusively for Pearson by Dr. Conley, the Conley Readiness Index assesses mastery in each of the "Four Keys" that are critical to college and career readiness:

KEY COGNITIVE STRATEGIES	KEY CONTENT KNOWLEDGE	KEY LEARNING SKILLS & TECHNIQUES	KEY TRANSITION KNOWLEDGE & SKILLS
Think	**Know**	**Act**	**Go**
Problem Formulation Hypothesize Strategize	**Structure of Knowledge** Key Terms and Terminology Factual Information Linking Ideas Organizing Concepts	**Ownership of Learning** Goal Setting Persistence Self-awareness Motivation Help-seeking Progress Monitoring Self-efficacy	**Contextual** Aspirations Norms/Culture
Research Identify Collect	**Attitudes Toward Learning Content** Challenge Level Value Attribution Effort	**Learning Techniques** Time Management Test Taking Skills Note Taking Skills Memorization/recall Strategic Reading Collaborative Learning Technology	**Procedural** Institution Choice Admission Process
Interpretation Analyze Evaluate			**Financial** Tuition Financial Aid
Communication Organize Construct	**Technical Knowledge & Skills** Specific College and Career Readiness Standards		**Cultural** Postsecondary Norms
Precision & Accuracy Monitor Confirm			**Personal** Self-advocacy and Institutional Context

Topics include:

Student Success Learning Path

- Conley Readiness Index
- College Transition
- Communication
- Creating an Academic Plan
- Critical Thinking
- Financial Literacy
- Goal Setting
- Information Literacy
- Learning Preferences
- Listening and Note Taking
- Majors and Careers Exploration
- Memory and Studying
- Online Learning
- Problem Solving
- Reading and Annotating
- Stress Management
- Test Taking
- Time Management

Career Success Learning Path

- Career Portfolio
- Interviewing
- Job Search
- Self-Management Skills at Work
- Teamwork
- Workplace Communication
- Workplace Etiquette

Assessment

Beyond the Pre- and Post-Full Course Diagnostic Assessments and Pre- and Post-Tests within each module, additional learning-outcome-based tests can be created using a secure testing engine, and may be printed or delivered online. These tests can be customized by editing individual questions or entire tests.

Reporting

Measurement matters—and is ongoing in nature. MyStudentSuccessLab lets you determine what data you need, set up your course accordingly, and collect data via reports. The high quality and volume of test questions allows for data comparison and measurement.

Content and Functionality Training

The Instructor Implementation Guide provides grading rubrics, suggestions for video use, and more to save time on course prep. Our Best Practices Guide and "How do I…" YouTube videos indicate how to use MyStudentSuccessLab, from getting started to utilizing the Gradebook.

Peer Support

The Student Success Community site is a place for you to connect with other educators to exchange ideas and advice on courses, content, and MyStudentSuccessLab. The site is filled with timely articles, discussions, video posts, and more. Join, share, and be inspired!
www.mystudentsuccesscommunity.com

The Faculty Advisor Network is Pearson's peer-to-peer mentoring program in which experienced MyStudentSuccessLab users share best practices and expertise. Our Faculty Advisors are experienced in one-on-one phone and email coaching, presentations, and live training sessions.

Integration and Compliance

You can integrate our digital solutions with your learning management system in a variety of ways. For more information, or if documentation is needed for ADA compliance, contact your local Pearson representative.

MyStudentSuccessLab users have access to:

- Full course Pre- and Post-Diagnostic Assessments linked to learning outcomes

- Pre- and Post-tests dedicated to individual topics

- Overviews that summarize objectives and skills

- Videos on key issues "by students, for students"

- Practice exercises that instill student confidence

- Graded activities to build critical-thinking and problem-solving skills

- Journal writing assignments with online rubrics for consistent, simpler grading

- Resources like Finish Strong 24/7 YouTube videos, calculators, professionalism/research & writing/ student success tools

- Student inventories including **Conley Readiness Index** and **Golden Personality**

Students utilizing MyStudentSuccessLab may purchase Pearson texts in a number of cost-saving formats— including eTexts, loose-leaf Books à la Carte editions, and more.

CourseConnect™
Trust that your online course is the best in its class.

Designed by subject matter experts and credentialed instructional designers, CourseConnect offers award-winning customizable online courses that help students build skills for ongoing personal and professional development.

CourseConnect uses topic-based, interactive modules that follow a consistent learning path—from introduction, to presentation, to activity, to review. Its built-in tools—including user-specific pacing charts, personalized study guides, and interactive exercises—provide a student-centric learning experience that minimizes distractions and helps students stay on track and complete the course successfully. Features such as relevant video, audio, and activities, personalized (or editable) syllabi, discussion forum topics and questions, assignments, and quizzes are all easily accessible. CourseConnect is available in a variety of learning management systems and accommodates various term lengths as well as self-paced study. And, our compact textbook editions align to CourseConnect course outcomes.

Choose from the following three course outlines ("Lesson Plans")

Student Success

- Goal Setting, Values, and Motivation
- Time Management
- Financial Literacy
- Creative Thinking, Critical Thinking, and Problem Solving
- Learning Preferences
- Listening and Note-Taking in Class
- Reading and Annotating
- Studying, Memory, and Test-Taking
- Communicating and Teamwork
- Information Literacy
- Staying Balanced: Stress Management
- Career Exploration

Career Success

- Planning Your Career Search
- Knowing Yourself: Explore the Right Career Path
- Knowing the Market: Find Your Career Match
- Preparing Yourself: Gain Skills and Experience Now
- Networking
- Targeting Your Search: Locate Positions, Ready Yourself
- Building a Portfolio: Your Resume and Beyond
- Preparing for Your Interview
- Giving a Great Interview
- Negotiating Job Offers, Ensuring Future Success

Professional Success

- Introducing Professionalism
- Workplace Goal Setting
- Workplace Ethics and Your Career
- Workplace Time Management
- Interpersonal Skills at Work
- Workplace Conflict Management
- Workplace Communications: Email and Presentations
- Effective Workplace Meetings
- Workplace Teams
- Customer Focus and You
- Understanding Human Resources
- Managing Career Growth and Change

Custom Services
Personalize instruction to best facilitate learning.

As the industry leader in custom publishing, we are committed to meeting your instructional needs by offering flexible and creative choices for course materials that will maximize learning and student engagement.

Pearson Custom Library

Using our online book-building system, create a custom book by selecting content from our course-specific collections that consist of chapters from Pearson Student Success and Career Development titles and carefully selected, copyright-cleared, third-party content and pedagogy.
www.pearsoncustomlibrary.com

Custom Publications

In partnership with your Pearson representative, modify, adapt, and combine existing Pearson books by choosing content from across the curriculum and organizing it around your learning outcomes. As an alternative, you can work with your Editor to develop your original material and create a textbook that meets your course goals.

Custom Technology Solutions

Work with Pearson's trained professionals, in a truly consultative process, to create engaging learning solutions. From interactive learning tools, to eTexts, to custom websites and portals, we'll help you simplify your life as an instructor.

Online Education

Pearson offers online course content for online classes and hybrid courses. This online content can also be used to enhance traditional classroom courses. Our award-winning CourseConnect includes a fully developed syllabus, media-rich lecture presentations, audio lectures, a wide variety of assessments, discussion board questions, and a strong instructor resource package.

For more information on custom Student Success services, please visit www.pearsonlearningsolutions.com.

1

Attitude, Goal Setting, and Life Management

future • dreams • happiness

After studying these topics, you will benefit by:

- Discovering the influence professionalism and positive human relations have on personal, academic, and career success
- Knowing how individual personality, attitude, and values affect the workplace
- Recognizing how self-efficacy and personal branding affect your confidence
- Developing a strategy to deal with past negative experiences and other barriers to success
- Examining the impact goal setting has on creating a life plan in today's economy
- Choosing priorities to support your goals

HOW DO YOU RATE?

Are you self-centered?	Yes	No
1. Do you rarely use the word I in conversations?	☐	☐
2. When in line with coworkers, do you let coworkers go ahead of you?	☐	☐
3. Do you keep personal work accomplishments private?	☐	☐
4. Do you rarely interrupt conversations?	☐	☐
5. Do you celebrate special events (e.g., birthdays, holidays) with your coworkers by sending them a card, a note, or small gift?	☐	☐

▶ If you answered "yes" to two or more of these questions, well done. Your actions are more focused on the needs of others and you are most likely not self-centered.

All About You

Congratulations! You are about to embark on a self-discovery to identify how to become and remain productive and successful in the workplace. The first step in this self-discovery is to perform a simple exercise. Look in a mirror and identify the first three words that immediately come to mind.

These three words are your mirror words. **Mirror words** describe how you view yourself and how you believe others view you. Your perception of yourself influences your relationship with coworkers and your workplace performance.

This text is all about professionalism in the workplace. The goal of both your instructor and the authors is to not only assist you in securing the job of your dreams, but to keep that great job and advance your career based on healthy, quality, and productive work habits that benefit you, your coworkers, and your organization. **Professionalism** is defined as workplace behaviors that result in positive business relationships. This text provides you tools to help you experience a more fulfilling and productive career. The secret to healthy relationships at work is to first understand yourself. Once you understand your personal needs, motivators, and irritants, it becomes easier to understand and successfully work with others. This is why the first part of this chapter focuses on your personality, your values, and your self-concept.

An individual's personality and attitude dictate how he or she responds to conflict, crisis, and other typical workplace situations. Each of these typical workplace situations involves working with and through people. Understanding your own personality and attitude makes it much easier to understand your reaction to others' personalities and attitudes.

Human relations are the interactions that occur with and through people. These interactions create relationships. Therefore, you theoretically have relationships with everyone you come into contact with at work. For an organization to be profitable, its employees must be productive. It is difficult to be productive if you cannot work with your colleagues, bosses, vendors, and/or customers. Workplace productivity is a result of positive workplace interactions and relationships.

Personality is a result of influences, and there are many outside influences that affect workplace relationships. These influences may include immediate family, friends, extended family, religious affiliation, and even society as a whole. Conversely, experiences and influences at work affect your personal life. Therefore, to understand workplace relationships, you must first understand yourself.

Personality and Values

Behavior is a reflection of personality. **Personality** is a stable set of traits that assist in explaining and predicting an individual's behavior. Personality traits can be positive, such as being caring, considerate, organized, enthusiastic, or reliable. However, personality traits can also be negative, such as being rude, unfocused, lazy, or immature. For example, if you are typically organized at work and suddenly you become disorganized, others may believe something is wrong because your disorganized behavior is not in sync with your stable set of organized traits. An individual's personality is shaped by many variables, including past experience, family, friends, religion, and societal influences. Perhaps a family member was incredibly organized and passed this trait on to you. Maybe someone in your sphere of influence was incredibly disorganized, which influenced you to be very organized. These experiences (positive or not) shape your values. **Values** are things that are important to you as an individual based on your personal experiences and influences. These influences include religion, family, and societal issues such as sexual preference, political affiliation, and materialism. Note that you may have good or bad values. You may value achievement, family, money, security, or freedom. For example, one individual may not value money because he or she has been told that "money is the root of all evil." Contrast this with an individual who values money because he or she has been taught that money is a valuable resource used to ensure a safe, secure future. Because values are things that are important to you, they will directly affect your personality. If you have been taught that money is a valuable resource, you may be very careful in your spending. Your personality trait will be that of a diligent, hardworking person who spends cautiously. A more in-depth discussion of values and how they relate to business ethics is presented in Chapter 5.

Topic Situation

While in school, Charley worked hard to secure a new job as an assistant at his college bookstore. Charley's parents are both college graduates with successful careers, which influences Charley's values and beliefs in the ability to perform successfully at school and work. However, many of Charley's friends are not attending college, and have a hard time securing and/or maintaining employment. For this reason, Charley gets no support from these friends regarding earning a degree and holding a job.

TOPIC RESPONSE

If Charley continues to associate with his non-supportive friends, how could these friendships influence Charley's performance at school and work?

Attitude

An **attitude** is a strong belief about people, things, and situations. For example, you either care or do not care how your classmates feel about you. Your attitude is related to your values and personality and affected by past success and failures. Using the previous example, if you value money, your attitude will be positive toward work because you value what you get in return for your work effort—a paycheck. Attitude affects performance: An individual's performance significantly influences a group's performance, and a group's performance, in turn, affects an organization's performance. Think about a barrel of juicy red apples. Place one bad apple in the barrel of good apples, and, over time, the entire barrel will be spoiled. That is why it is so important to evaluate personal influences. The barrel reflects your personal goals and your workplace behavior. Your attitude affects not only your performance, but also the performance of those with whom you come in contact.

Does this mean you avoid anyone you believe is a bad influence? Not necessarily. You cannot avoid certain individuals, such as relatives and coworkers. However, you should be aware of the impact individuals have on your life. If certain individuals have a negative influence, avoid or limit your exposure to them (bad apple). If you continue to expose yourself to negative influences, you can lose sight of your goals, which may result in a poor attitude and poor performance. Choose your friendships wisely and surround yourself with positive people. Positive people are truthful, faithful, loving, and supportive. Negative people interfere with you reaching your goals by making you uncomfortable or by distracting you.

Think About It

Identify one friend that you believe is a positive influence on you and a friend that is a negative influence. How should you handle these relationships?

Talk It Out

What cartoon character best reflects your personality and why?

Self-Efficacy and Its Influences

Review your "mirror words" from the beginning of this chapter. Were your words positive or negative? Whatever you are feeling is a result of your **self-concept**. Self-concept is how you view yourself. Thinking you are intelligent or believing you are attractive are examples of self-concept. **Self-image** is your belief of how others view you. If your self-concept is positive and strong, you will display confidence and not worry about how others view you and your actions. If you are insecure, you will rely heavily on what others think of you. Although it is important to show concern for what others think of you, it is more important to have a positive self-concept. Note that there is a difference between being conceited and self-confident. Those who behave in a conceited manner have too high an opinion of themselves as compared to others. People are drawn to individuals who are humble, display a good attitude, are confident, and are consistently positive. It is easy to see the tremendous impact both personality and attitude have in the development of your self-esteem and self-concept. One final factor that influences self-concept and performance is that of self-efficacy. **Self-efficacy** is your belief in your ability to perform a task. For example, if you are confident in your math abilities, you will most likely score high on a math exam because you believe you are strong in that subject. However, if you are required to take a math placement exam for a job and are not confident in your math abilities, you will most likely not perform well. The way you feel about yourself and your environment is reflected in how you treat others. This is called **projection**. A positive self-concept will be projected toward others.

Envision a hand mirror. The handle of the mirror (the foundation) is your personality. The frame of the mirror represents your personal values. The mirror itself is your attitude, which is reflected for you and the world to see. The way you view yourself is your self-concept; the way you believe others see you is your self-image. As you begin networking with others, interviewing for a new job, or embarking on a new career, create a **personal brand**. A personal brand reflects traits you want others to think of when they think of you. These personal traits may include your appearance, your values, or specific knowledge or skills that make you unique, interesting, and of value to others. Throughout this text, you will be gaining additional tools designed to improve your professionalism. Use these tools to refine your personal brand and make a commitment to continue enhancing your brand. Doing so will contribute to a positive self-concept and increase your odds for both personal and professional success.

Exercise 1.1

Define your personal brand. Identify desired appearance, personality, knowledge and skills, personal values, and attitude.

_____

_____

_____

Dealing with Negative Baggage

Many of us have experienced people who appear to have a chip on their shoulder that negatively influences their behavior. The negativity is reflected in an individual's personality. More often than not, the "chip" is a reflection of a painful past experience. What many do not realize is that negative past experiences sometimes turn into personal baggage that creates barriers to career success. Examples of negative past experiences may include traumatic issues such as an unplanned pregnancy or a criminal offense. Other times, the negative experience involved a poor choice or a failure at something that had great meaning. These experiences are the ones that most heavily influence one's personality, values, and self-concept, and in turn, may affect workplace attitude and performance.

Topic Situation

When starting high school, Keira made a poor choice and got in minor trouble with the law. Keira paid her dues, yet is still embarrassed and sometimes feels unworthy of a successful future. Keira is trying to climb the mountain of success carrying a hundred-pound suitcase. The suitcase is filled with the thoughts of a previous poor choice and embarrassment. Because of Keira's motivation to complete college, most friends and acquaintances are unaware of her past mistake. However, if Keira continues to carry this negative baggage, she may lose sight of her goals.

TOPIC RESPONSE

What steps should Keira take to help her achieve her goals?

If you have had a negative experience that is hindering your ability to succeed, recognize the impact your past has on your future. Although you cannot change yesterday, you can most certainly improve your today and your future. Take these steps toward a more productive future:

1. *Confront your past.* Whatever skeleton is in your past, admit that the negative event occurred. Do not try to hide or deny that it happened. There is no need to share the episode with everyone, but it may help to confidentially share the experience with someone you trust (friend or trained professional) who had no involvement with the negative experience. Acknowledgement of the negative event is the first step toward healing.
2. *Practice forgiveness.* Past negative experiences hurt. A process in healing is to forgive whoever hurt you. Forgiveness does not justify that what occurred was acceptable, but reconciles in your heart that you are dealing with the experience and are beginning to heal. Identify who needs forgiveness. The act of forgiveness may involve a conversation with someone, or it may just involve you deciding to no longer carry this burden.
3. *Move forward.* Let go of hurt, guilt, and/or embarrassment. Do not keep dwelling on the past and using it as an excuse or barrier toward achieving your goals. If you are caught in this step, physically write the experience down on a piece of paper and the words "I forgive Joe" (replace the name with the individual who harmed you). Then take the paper and destroy it. This physical act puts you in control and allows you to visualize the negative experience being diminished. As you become more confident in yourself, your negative experience becomes enveloped with the rest of your past and frees you to create a positive future.

This sometimes painful process is necessary if your goal is to become the best individual you can be. Dealing with negative baggage is not something that happens overnight. As mentioned previously, some individuals may need professional assistance to help them through the process. There is no shame in seeking help. In fact, there is great freedom when you have finally let go of the baggage and are able to climb to the top of the mountain unencumbered.

Locus of Control

The reality is that you will not always be surrounded by positive influences and you cannot control everything that happens in your life. Your attitude is affected by who you believe has control over situations that occur in your life, both personally and professionally. The **locus of control** identifies who you believe controls your future. An individual with an *internal* locus of control believes that he or she controls his or her own future. An individual with an *external* locus of control believes that others control his or her future.

Extremes on either end of the locus of control are not healthy. Realize that individual effort and a belief in the ability to perform well translate to individual success. External factors also influence your ability to achieve personal goals. You cannot totally control the environment and future. Power, politics, and other factors discussed later in the text play an important part in the attainment of goals. Successful individuals take personal responsibility and avoid blaming others.

Learning Styles

Another element of personality is one's **learning style.** Learning styles define the method of how you best take in information and/or learn new ideas. There are three primary learning styles: visual, auditory, and tactile/kinesthetic.

To determine what your dominant learning style is, perform this simple exercise. Imagine you are lost and need directions. Do you:

   a. Want to see a map
   b. Want someone to tell you the directions
   c. Want to draw or write down the directions yourself

If you prefer answer *a,* you are a visual learner. You prefer learning by seeing. If you selected *b,* you are an auditory learner. You learn best by hearing. If you selected *c,* you are a tactile/kinesthetic learner, which means you learn best by feeling, touching, or holding. No one learning style is better than the other. However, it is important to recognize your primary and secondary learning styles so that you can get the most out of your world (in and out of the classroom or on the job). As a visual learner, you may digest material best by reading and researching. Auditory learners pay close attention to course lectures and class discussions. Tactile/kinesthetic learners will learn best by performing application exercises and physically writing course notes. Recognize what works best for you and implement that method to maximize your learning experience. Also recognize that not everyone learns the same way you do and not all information is presented in your preferred method. With that recognition, you can become a better classmate, team member, coworker, and boss.

Copyright © 2016, 2013, 2011 by Pearson Education, Inc.

Exercise 1.2

Apply the learning styles discussed and complete the following statements.

In the classroom, I learn best by _____

In the classroom, I have difficulty learning when _____

How will I use this information to perform better? _____

Importance and Influences of Personal Goal Setting

Everyone has dreams. These dreams may be for a college degree, a better life for loved ones, financial security, or the acquisition of material items such as a new car or home. Goal setting is the first step toward turning a dream into a reality. This important process provides focus and identifies specific steps that need to be accomplished. It is also a common practice used by successful individuals and organizations. A **goal** is a target. Think of a goal as a reward at the top of a ladder. To reach a goal, you need to progress up each step of the ladder. Each step contributes to the achievement of a goal and supports your personal values. Goals help you decide what you want in your future, increase self-concept, and help overcome procrastination, fear, and failure.

When you set goals, career plans become more clear and meaningful. They motivate you to continue working to improve yourself and help you achieve—not just hope for—what you want in life.

Topic Situation

At 22 years of age, Austin had only a high-school education. After working odd jobs at minimum wage since graduating from high school, Austin decided to attend college to become a Certified Public Accountant (CPA). Austin set a long-term goal to finish college in five years. Self-supporting and having to work, he set a realistic goal to obtain an associate degree in accounting within three years. After achieving that goal, Austin plans to find a job as an account clerk while finishing school. This goal will increase his income and self-confidence. Still committed to becoming a CPA, he plans to earn a bachelor's degree in accounting within two years after receiving the associate degree.

TOPIC RESPONSE

What are specific steps Austin can take to ensure he reaches his goal of becoming a CPA?

Goals can and should be set in all major areas of your life, including personal, career, financial, educational, and physical. Goals help maintain a positive outlook. They also contribute to creating a more positive perception of yourself and result in improved human relations with others.

Example of Austin's goals:

Talk It Out

Share one goal you have set for this class.

Five-year long-term goal	Obtain a bachelor's degree in accounting
Three-year long-term goal	Obtain an associate degree in accounting and secure a job as an account clerk
One-year short-term goal	Successfully pass the appropriate courses toward the associate degree and identify an internship
Now	Apply for school and find a part-time job to obtain work experience

How to Set Goals

As explained earlier, achieving goals is like climbing a ladder. Imagine that there is a major prize (what you value most) at the top of the ladder. The prize can be considered your long-term goal, and each step on the ladder is a progressive short-term goal that helps you reach the major prize.

Set short-term and long-term goals and put them in writing. **Long-term goals** are goals that take longer than a year to accomplish, with a realistic window of up to 10 years.

To set a goal, first identify what you want to accomplish in life. Write down everything you can think of, including personal, career, and educational dreams. Next, review the list and choose which items you most value. In reviewing your list, ask yourself where you want to be in one year, five years, and 10 years. The items you identified are your long-term goals. Keep each goal realistic and something you truly want. Each goal should be attainable, yet challenging enough to work toward. Identify why each goal is important to you. This is a key step toward setting yourself up for success. Next, identify opportunities and potential barriers toward reaching these goals. Remember Austin's goal to be a CPA? Austin believes becoming a CPA represents success. It is important to him, and it is a realistic goal that can be reached.

Exercise 1.3

Identify educational, personal, and professional accomplishments you would like to achieve in 5–10 years.

_____

_____

_____

Short-term goals are goals that can be reached within a year's time. They are commonly set to help reach long-term goals. Businesses often refer to short-term goals as objectives because they are measurable and have a one year or less time line. Short-term goals can be achieved in one day, a week, a month, or even several months. As short-term goals are met, long-term goals should be updated. Just like long-term goals, short-term goals (objectives) must be realistic, achievable, and important to you. They need to be measurable so you know when you have actually reached them.

An additional long-term goal for Austin is to buy a car one year after graduation. Austin has set several short-term goals, one being to save a specific amount of money each month. To do this, he needs to work a certain number of hours each week. He also needs to be specific about the type of car, whether to buy used or new, and whether he needs to take out a loan. The answers to these questions will determine how much money Austin will need to save each month and if the one-year time frame is realistic.

Exercise 1-4

Using the goals you identified in Exercise 1.3, identify how you can turn each dream into a reality.

_____

_____

_____

A popular and easy goal-setting tool is the SMART method. **SMART** is an acronym for writing goals to ensure they are specific, measurable, achievable, relevant, and time based.

			Example
S	Specific	Clearly identify what exactly you want to accomplish and, if possible, make your goal quantifiable. This makes your goal specific.	Become a manager for a top accounting firm
M	Measurable	Make your goal measurable. Identify how you will know when you have achieved your goal.	Having the job as a manager
A	Achievable	Keep your goal achievable but not too easily attainable or too far out of reach. A good achievable goal is challenging, yet attainable and realistic.	Getting good grades in college and gaining work experience along the way
R	Relevant	Relevant personal goals have meaning to its owner. The goal should belong to you, and you should have (or have access to) the appropriate resources to accomplish the goal.	I want to do this
T	Time Based	Attach a specific date or time period to provide a time frame for achieving the goal.	By 2021

For example, instead of writing, "I will become a manager in the future," write, "After attending college and getting work experience, I will become a manager with a top accounting firm by the beginning of the year 2021." After you have written a goal, give it the SMART test to increase its probability for success.

After you have written positive and detailed goals, there are a few additional aspects of goal setting to consider. These include owning and taking control of your goals.

Owning the goal ensures that the goal belongs to you. You are the one who should decide your goals, not your parents, spouse, significant other, friends, relatives, or anyone else. For example, if Austin goes to college because it is his personal dream to be a CPA, that goal will be accomplished. However, if Austin becomes a CPA because his parents want him to be a CPA, this will not be Austin's goal and it will be harder to accomplish.

Take control of your goal by securing information necessary to accomplish it. Know what resources and constraints are involved, including how you will use resources and/or get around constraints. If your goal is related to a specific career, identify what is required to attain that career in regard to education, finances, and other matters. Clarify the time needed to reach your goals by writing them as short-term or long-term goals. Applying the concept of locus of control, remember that not every factor is within your control. Therefore, be flexible and realistic with your goals and the time you take to achieve them.

Creating a Life Plan

Identifying goals contribute to the creation of a **life plan**. A life plan is a written document that identifies goals for all areas of your life, including career, family/social, spiritual, and financial.

Consider what you want in the following areas of your life:

- *Education and career:* Degree attainment, advanced degrees, respectable job titles, specific employers
- *Social and spiritual:* Marriage, family, friends, religion
- *Financial:* Home ownership, car ownership, investments
- *Activities:* Travel, hobbies, life experiences

Create goals for each of these major life areas and establish goals that reflect your values. Note that some goals may blend into two or more areas. Remember that goals can change over time; stay focused but flexible.

It is common for younger students to be uncertain of their career goals. Others may feel overwhelmed that they have a life goal but lack the necessary resources to accomplish one or more goals. Education is an important key to achieving personal and career goals and no one can take your knowledge from you. When writing your life goals, consider the degrees/certificates, the time frame, the financial resources, and the support network you will require for educational success. Make college course choices based on your desired educational goals. Choose courses that will benefit you, help you explore new concepts, and challenge you. To be successful in your career, it is important to enjoy what you do. Select a career that supports your short-term and long-term goals. When planning your career consider:

- Why your target career is important to you
- What resources are needed to achieve your career goal
- How you will know you have achieved career success

People choose careers for different reasons, including earning power, status, intellect, values, and self-satisfaction. If there is a career center available at your college, take time to visit and explore the various resources it offers. There are also several personality and career interest tests that will help you determine career options. Career assessments are offered at many college career centers and online. These useful assessments assist in identifying interests, abilities, and personality traits to determine which career will suit you best. Take advantage of all available resources and gather information to assist you in making the optimal career decision. Conduct an Internet search, perform an internship, volunteer, interview, or job shadow someone working in the field that interests you. Doing so will help clarify your goals and life plan. An additional discussion on career exploration is presented in a later chapter.

Web Search

Discover your personality: Take one of the personality and career assessments available on MyStudent-SuccessLab, or conduct a web search to identify an online quiz that will help you discover your personality and career interests.

Consider the type of personal relationships you want in the future. Goals should reflect your choice of marriage, family, friends, and religion. Identify where you want to be financially. Many people dream of becoming a millionaire, but you need to be realistic. Think about what kind of house you want to live in and what type of car you want to drive. If a spouse and children are in your future, account for their financial needs as well. Also identify what outside activities you enjoy, including hobbies and travel. The personal financial plan you create will be a part of achieving these goals. You will work on your personal financial plan in greater detail in the next chapter. Think about what results and rewards will come from achieving your goals.

Intrinsic and extrinsic rewards motivate individuals to achieve their goals. **Intrinsic rewards** come from within you and reflect what you value, including such things as self-satisfaction and pride of accomplishment. **Extrinsic rewards** come from external sources and include such things as money and praise. Identify what type of intrinsic and extrinsic rewards motivate you, and then use them to help you maintain a positive outlook while working toward your goals.

Talk It Out

Share common rewards that are important to you. Identify these rewards as intrinsic or extrinsic.

Priorities

Priorities determine what needs to be done and in what order. Properly managing priorities is a valuable tool for reaching goals. Not only is prioritizing important in your personal life, but it will be necessary at work.

As you work toward your goals, priorities may change. There may be a period when your first priority is not necessarily what is most important in life; it is just that a particular activity demands the most attention at that specific point in time. For example, if Amelia has a young child, that child is important in Amelia's life. However, if Amelia is attending college and needs an evening to study for a big exam, the priority is to study for the exam. That does not mean the exam is more important than the child. However, passing the exam is a step toward a better future for Amelia and her child.

Amelia's decision is called a **trade-off**. A trade-off is giving up one thing to do something else. Another example involving Amelia is her decision to purchase a car in one year; she needs to save a certain amount of money each month. In order to do this, she may have to give up buying coffee each morning and instead make her coffee at home in order to save enough money to purchase the car.

Life plans require flexibility. When working toward goals, be flexible. Times change, technology changes, and priorities may change, all of which affect your goals. Reevaluate goals at least once a year. You may need to update or revise your goals and/or time lines more frequently than once a year because a situation changed. Do not abandon a goal because a situation changed—simply modify the goal and move forward.

Talk It Out

Identify priorities and trade-offs for successfully completing this course.

Your Personal Handbook

This book is designed as a personal handbook that leads you on an exciting journey toward creating both personal and career plans. On this journey you will develop a respect and understanding of basic personal financial management and the influence finances have on many areas of your life. Self-management skills, including time, stress, and organization, will be addressed, as well as professional etiquette and dress. Workplace politics, their implications for performance, and how to successfully use these politics in your favor will be discussed, as will your rights as an employee. These newfound workplace skills will improve your ability to lead, motivate, and successfully work with others in a team setting. Finally, you will learn how to handle conflict and work with difficult coworkers.

As we move through key concepts in this text, begin developing a positive attitude and believe in yourself and your abilities. Equally important is that you learn from your past. Little by little, you will make lifestyle changes that will result in you being a better individual, which will result in you becoming an even better employee. It all translates to success at work and success in life.

MyStudentSuccessLab   Please visit **MyStudentSuccessLab**: Anderson|Bolt, Professionalism Skills for Workplace Success, 4/e for additional activities, resources, and outcomes assessments.

Workplace Dos and Don'ts

Do realize the impact your personality has on overall workplace performance	*Don't* assume that everyone thinks and behaves like you
Do believe that you are a talented, capable human being	*Don't* become obsessed with how others view you
Do let go of past baggage	*Don't* keep telling everyone about a past negative experience
Do set goals in writing	*Don't* set goals that are too difficult to reach
Do set long-term and short-term goals	*Don't* give up on goals
Do make your goals attainable	*Don't* wait to create goals
Do have measurable goals	*Don't* create unrealistic goals
Do set priorities and include trade-offs and flexibility when setting goals	*Don't* give up when working to reach your goals

Concept Review and Application

You are a Successful Student if you:

- Explain the importance of professionalism
- Create a strategy to enhance your personal brand
- Write a life plan

Summary of Key Concepts

- How you view yourself dictates how you treat others and what type of employee you will be.

- Your views of yourself, your environment, and your past experiences comprise your personality, values, attitude, and self-efficacy.

- Negative past experiences create unnecessary baggage that either delays or prevents you from reaching your goals. Acknowledge and begin dealing with these negative experiences.

- There are three primary learning styles: visual, auditory, and tactile/kinesthetic (sight, sound, and touch). Individuals must recognize how they best learn and also be aware that others may or may not share their same learning style.

- Goal setting is important in helping you keep focused. It will enhance your self-concept and help you become more successful in all areas of your life.

- As goals are reached, motivation and self-confidence will increase.

- Goals need to be put into writing. They need to be realistic and measurable. Know who owns the goals and who controls the goals. A time frame is needed to know when you plan on reaching these goals.

- Long-term goals are set to be achieved in five to 10 years.

- Short-term goals are achieved within a year's time and are needed to reach long-term goals.

- When creating a life plan, consider all aspects of your life, including personal, career, and education.

- Flexibility and properly managing priorities are needed to successfully achieve goals.

Self-Quiz MATCHING KEY TERMS

Match the key term to the definition using the identifying number.

Key Terms	Answer	Definitions
Attitude		1. Your belief in your ability to perform a task
Extrinsic rewards		2. Identifies who you believe controls your future
Goal		3. An individual's perception of himself or herself
Human relations		4. A strong belief about people, things, and situations
Intrinsic rewards		5. Giving up one thing to do something else
Learning style		6. An individual's perception of how others view him or her
Life plan		7. A target in your life plan
Locus of control		8. How you best take in new information and/or learn new ideas
Long-term goals		9. Describe how you view yourself and how you believe others view you
Mirror words		10. Rewards that come from within and may include self-satisfaction and pride of accomplishment
Personal brand		11. Determine what needs to be done and in what order
Personality		12. Things that are important to an individual
Priorities		13. Rewards from external sources such as money and praise
Professionalism		14. Workplace behaviors that result in positive business relationships
Projection		15. Interactions occurring with and through people
Self-concept		16. A target that takes longer than one year to accomplish
Self-efficacy		17. The way you feel about yourself is reflected in how you treat others
Self-image		18. Set of traits that assist in explaining and predicting an individual's behavior
Short-term goals		19. Reflects traits you want others to think of when they think of you
Smart		20. Goals that can be reached within a year's time
Trade-off		21. Acronym for goal setting method
Values		22. A written document that identifies goals in all areas of your life

Think Like a Boss

1. How would you deal with an employee who displays poor self-efficacy?

2. How would recognizing different learning styles help you be a better boss?

3. Why is it important that an employer ensure that employees set personal and career goals?

Activities

Activity 1.1

Write four words to describe your ideal self-concept (personal brand).

1	3
2	4

What steps are necessary to make your desired personal brand a reality?

_____

_____

Activity 1.2

What factors affect your attitude toward educational success?

Positive Factors	Negative Factors

Activity 1.3

Identify and write your long-term personal, educational, and career goals, giving each the SMART test.

Personal	Education	Career

Activity 1.4

Write three short-term goals to support each long-term goal identified in Activity 1.3, giving each the SMART test.

	Personal	Education	Career
Long-term goal			
Short-term goal 1			
Short-term goal 2			
Short-term goal 3			

2
Personal Financial Management

security • independence • choices

After studying these topics, you will benefit by:

- Recognizing the significance of money management and budgeting on personal and professional success
- Distinguishing the wise use of credit and identifying debt management resources
- Evaluating alternatives for financing your education through student loans, financial aid, and other resources
- Discovering the impact your credit report has on your financial future
- Identifying methods to protect yourself from identity theft
- Assessing money wasters, emotional spending, and the impact money has on relationships

HOW DO YOU RATE?

How personal are you making finance?	Yes	No
1. I have a personal budget.	☐	☐
2. I routinely use my personal budget.	☐	☐
3. I can tell you how much money I have in my checking and/or savings account(s).	☐	☐
4. I can tell you how much I currently owe in long-term debt/credit card bills.	☐	☐
5. I routinely pay off the entire balance of my monthly credit card bill.	☐	☐

▶ If you answered "no" to two or more of these questions, treat this as an opportunity to get personal with your finances. Knowing and applying personal financial concepts will enable you to achieve personal goals, improve your self-concept, and better understand business.

Financial Management

We go to work to earn money. What we do with our money is based on our goals and values. It is difficult to be productive and sometimes trusted at work if an individual does not have his or her personal financial affairs in order. Individuals without a personal financial plan usually have little control over spending, which may result in stress and financial crisis. Creating and utilizing a personal financial plan helps form habits that contribute to the realization of long-term goals and professional success. **Personal financial management** is the process of controlling personal income and expenses. **Income** is money coming in. This money may come from parents, grants, scholarships, student loans, and/or a job. While you are a student, your income may be minimal. However, after finishing college, you will begin a new career with (ideally) an increased income. As your income increases your expenses will also increase. Learning to properly handle your money now will make managing your money easier as your income grows.

An **expense** is money going out, or spent. Examples of student expenses include tuition, textbooks, school supplies, housing, and transportation. Basic life expenses include food, shelter, and clothing. Other common expenses are hobbies, entertainment, health care, and loans.

Personal Finances Affect Work Performance

Personal finance affects all areas of life. Finances are important in helping reach the goals you identified in Chapter 1. Personal financial management does not have to restrict your activities. Instead, it is a way to make your financial resources assist you in reaching goals while ensuring a healthy financial future. Now is the time to start making your money work for you by creating a personal financial plan.

Proper financial management includes monitoring your money and keeping your debt under control. It also involves maintaining a favorable credit report by using credit wisely and beginning savings and investment plans. Finally, it is wise to take steps to protect yourself from identity theft.

As explained at the start of this chapter, a lack of personal financial management can negatively affect your work situation. If you are not properly managing your finances, you will eventually have difficulty making purchases and paying

Topic Situation

TOPIC RESPONSE

Where are you unnecessarily spending money?

A coworker has been asking to borrow money for lunch from Oscar. Oscar has noticed that this coworker comes to work with a specialty coffee each morning and buys lunch every day. This is causing a strain on the relationship between Oscar and this coworker. Oscar has been on a strict budget since starting his job and instead of buying lunch he brings it from home. Oscar only goes to the coffee shop on special occasions. After loaning money to the coworker several times, Oscar decides to talk to his coworker. He shares with the coworker the importance of budgeting and helps the coworker create a budget of her own. After a while, the coworker stops asking Oscar and others for money. The coworker starts bringing lunch from home and treats herself to a specialty coffee only once in a while. A few months later the coworker thanks Oscar because she is now saving money.

your bills, leading to considerable stress. In turn, this stress will flow into the workplace and your performance will eventually deteriorate. Some employers require that you submit references and/or agree to a credit check prior to hiring, especially if your job requires working with money. Employers rationalize that if you cannot manage your personal finances, you may not be a responsible employee, nor can you be trusted with company resources.

Money Management

Many students struggle to keep up with expenses. Although you may currently have a job, your wages may not be very high. Perhaps you are low on cash and have just enough money to get through school.

Chapter 1 explained how to create goals. Many goals require time and money to achieve. Long-term and short-term financial goals are a necessary complement to personal goals. If you want to purchase a car in one year, you have to create a budget and save a specific amount of money in order to purchase the car when planned. Although it is sometimes tempting to spend money just because you have it, think about the future, practice self-control, and do not give in to the temptation to spend. Financial success begins with discipline and planning.

Web Search

Research money management apps or websites that will help keep track of your finances.

Personal Budgeting

The best way to manage money and still be able to buy some of the extras you want is to create a budget. A **budget** is a detailed financial plan used to allocate money for a specific time period. A budget reflects goals and identifies where your money will be spent in order to reach these goals. Control and prioritize your spending to match these goals. Be as precise and honest as you can when you are creating and working with a budget.

The first step in creating a budget is to identify goals. In Chapter 1, you created and identified goals for the future. Your written goals provide the foundation for your budget because your financial goals will be attached to your personal goals. The series of exercises and activities throughout this chapter assist you in developing these financial goals. To put your budget into action, determine your income and expenses. Remember, income is money coming in and an expense is money going out. It is best to establish a budget on a month-to-month basis; therefore, identify monthly income and expenses.

Start by determining all income that you receive on a monthly basis. If you know your income on a yearly basis, divide it by 12 (months) to identify monthly income. **Gross income** is the amount of money in a paycheck before paying taxes or other deductions. However, to make it easier to create your budget, use net income. **Net income** is the amount of money you have after all taxes and deductions are taken out of your gross pay.

After you have identified your monthly income, determine your expenses. Estimate how much you spend every month in each category. If you spend money in an area that is not listed, add another category to the list. Do not overuse the miscellaneous category. The idea is to track exactly where your money is being spent. Ideally, track all of your expenses over the next few months. Doing so will provide a true picture of where you really spend your money.

Fixed expenses are expenses that do not change from month to month, such as a monthly mortgage or rent payment. **Flexible expenses** (also referred to as *variable expenses*) are expenses that change from month to month, such as food or utilities. To identify monthly flexible expense amounts, take the past 12 months of that expense and use the average for your monthly budget.

Exercise 2.1

List everything you purchased in the last week (as much as you can remember). Then identify it as a want or a need.

_____

_____

_____

First budgets are rarely exact. Adjust your personal budget monthly as you track and identify specific income and expenses. An end-of-chapter activity provides the opportunity to create a personal budget. The following is an example of a budget.

	Estimated	Actual	Difference	Balance
	(what you think you will receive and spend for the month)	(actual amount you received or spent)	(actual amount minus estimated amount)	(income minus all expenses)
INCOME				
1. Net Income	$1,600	$1,760	$160	$1,760
EXPENSES				
2. Housing	500	500	0	1,260
3. Food	200	240	40	1,020

1. Your previous net income (take-home pay) has been $19,200 for the last year. To find your estimated monthly income you calculate $19,200 ÷ 12 (months in a year) = $1,600 per month. However, assume this month you got an unexpected raise to make your yearly net salary $21,120. Your new monthly salary would be $1,760. That would be your actual net income. The difference would be $1,760 − $1,600 = $160.
2. Your rent has been $500 a month (estimated). This is considered a fixed expense that would not change from month to month. The balance (income minus expenses) is now $1,760 − $500 = $1,260.
3. Your food usually costs about $200 a month (estimated). This month you ate out more and you spent $240 on food (actual). The difference is + $40. The balance is now $1,020.

The purpose of creating a budget is to determine your financial activities. By accurately keeping a budget, you may be surprised by how much you spend on certain items. Your budget will identify where you are spending money unnecessarily and will allow you to modify your spending while developing good personal financial management habits. A budget helps identify **money wasters**, which are small expenditures that you may not realize are consuming a larger portion of your income than expected. Common examples of money wasters include buying lunch instead of bringing it from home or buying soda from a machine instead of purchasing bulk at the store.

Here is a specific example of a money waster: On average, a specialty cup of coffee from a café is about $4.50. If you buy a cup of coffee five days a week, over a year's time you have spent $1,170 ($4.50 × 5 days × 52 weeks) on coffee. You may have been buying coffee because you do not have a coffeemaker and think you cannot afford to purchase one. Assume a specialty coffeemaker costs about $125 and, on average, flavored coffee runs about $12 a pound, and milk costs about $4. This will last one person at least two months. So the total spent for coffee for the year would be about $245 (($12 + $8) × 6 + $125). If you purchased the supplies and made your own coffee, you could save $925. Imagine what you could do with an extra $925 a year.

Think About It

How will a personal budget assist you with your finances?

Exercise 2.2

Without repeating the examples in this section, identify three common money wasters and alternatives that will save money.

_____

_____

_____

Cash management is not only the key to successful budgeting; it is also a means of avoiding money wasters. Carry only a small amount of cash, minimize using your ATM, and reduce trips to the ATM and/or bank. Most individuals spend more cash when it is readily available. A good cash management practice is to track every single transaction. Many individuals forget how much money has been spent and where it has gone. Physically record all deposits and withdrawals made with your ATM card, debit card, or checking account when they occur. Mentally keeping track of how much money you have in your account results in inaccurate accounting and overspending. Smartphones have apps designed to assist you in this practice. Prior to spending money, take time to seriously think about where your money is going and if the expense is necessary or if it is an impulse purchase. Implementing these cash management tips will help you discover when and where you are spending your money.

Debt Management

Debt is money owed. There is a difference between an expense and debt. As defined earlier, an expense is money going out. A common expense, like monthly rent, is a bill. A **loan** is a large debt that is repaid in smaller amounts over a period of time and has interest added to the payment. **Interest** is the cost of borrowing money and is the money you pay a lender for a loan. Debt includes all types of loans (car, home, school) and credit cards. You may already have some debt, such as a student loan or a credit card.

The use of credit cards is one way many people fall into debt. The inability to pay credit card debt causes individuals to file for bankruptcy, which results in long-term bad credit. Do not allow yourself to fall into a debt trap. The best way to avoid a debt trap is to purchase only what you can afford. The wise use of credit is discussed later in this chapter.

Attempt to maintain a positive net worth. **Net worth** is the amount of money that is yours after paying off debt. Net worth is determined by comparing your assets and liabilities. Personal **assets** are what you own. These are items that are worth money (for example, a car, home, and furniture). A **liability** is an obligation to pay what you owe. If you have a car loan, it is a liability.

Although your net worth may not currently be high, in the future, as you practice sound money management, your net worth will increase. You increase net worth by decreasing your liabilities and increasing personal assets.

Total assets − total liabilities = total net worth

If you are in debt, now is the time to begin getting yourself out. Seek advice and support from a parent, school counselor, or financial counselor. Talk with your creditor to work out a reduced payment or lower interest rate. Destroy but do not cancel unnecessary credit cards. For those in heavy debt, canceling credit cards has the potential to harm your credit score. Individuals in a debt hole still need credit.

Canceling or destroying a credit card does not eliminate the debt. You are obligated to pay all debt. Write down each credit debt, list the amount owed, the interest charged, and your monthly payment. Then prioritize your debt. Pay off the smallest amount owed or the amount with the highest interest first. After you have paid off one loan, apply the extra cash to the next debt on your priority list.

Do not ignore the warning signs of being in debt. Take action now. There are reputable national, non-profit credit-counseling services and credit-repair organizations that can assist you, such as the National Foundation for Consumer Credit or Myvesta Foundation.

Talk It Out

What are warning signs that you may be getting into debt?

Student Loans

When you created your life plan, securing a college degree was most likely one of your goals. Trying to identify how to pay for college without being strapped with a tremendous amount of debt upon graduation has become a big challenge for students. According to the College Board, approximately two-thirds of full-time undergraduate college students receive some type of financial aid in the form of grants, scholarships, loans, or work-study. A student loan is one of the first loans most students secure. Student loans are like all other loans, in that the borrower must pay back the loan with interest. As with any other loan, before you decide to secure a student loan, exhaust all other funding sources (if only to assist with partial tuition funding), including grants, scholarships, part-time work, personal savings, and family.

Should you determine that a student loan is necessary, identify the amount necessary and don't borrow more than that amount. When you secure your loan, only use these funds for direct school expenses. Refer to your life plan and focus on completing your degree within your time frame. In this case, time is money. The longer you take to secure your degree, the more likely you will rely on (and have to repay) student debt upon graduation. Although your school may include student loans as part of your application process, it is important to know what types of loans are offered.

There are two types of student loans: federal student loans and private student loans. In reviewing the following table, you will see that federal student loans have many advantages. However, some individuals do not qualify for a federal loan.

Federal Student Loans	Private Student Loans
• Funded by the government	• Made by private lender (such as a bank)
• Must be repaid after graduation, or when student drops below half-time status	• Paid while still in school
• Fixed interest rate and usually lower interest than a private loan	• Variable or fixed interest rate
• No credit check needed	• A credit check and/or a co-signer is required
• Interest may be tax deductible	• Interest may not be tax deductible
• Loans may be consolidated (into one loan)	• Cannot consolidate into a direct consolidation loan
• Payment may be tied to your income	• Prepayment penalties may exist
• No prepayment penalty fee	• Forgiveness programs are unlikely
• Offers loan forgiveness to those who are eligible	

To secure a federal student loan, you must first complete the Free Application for Federal Student Aid (FAFSA™). Search online for FAFSA.

Wise Use of Credit

The best way to stay out of a debt hole is to manage credit and loans and establish a savings plan. Many individuals receive credit or loans and make purchases they ultimately cannot afford.

When you are deciding on the best option for a loan or credit card, consider interest rates, hidden costs, the purpose of the loan or credit card, the amount of your payments, and how long it will take you to pay off the loan. Loan documents outline the lender's right to change the terms and conditions of a loan. Read and understand the fine print of loan documents prior to signing a loan agreement. Once you agree to the loan, you are legally obligated to abide by these terms and conditions.

As you increase your earning power, you may receive offers from credit card companies. Do not accept all credit offers. Credit is a privilege that should not be abused. The goal is to build and maintain good credit. Good credit aids in purchasing large items such as a car or home at a lower interest rate. Use a credit card as a tool to establish and maintain good credit. Spend wisely and pay off the balance each month. If you know you cannot pay the balance each month, do not use a credit card. Use the credit card only for items you can afford, and always make credit payments on time. Avoid taking cash advances against your credit card; credit companies charge higher interest rates for cash advances. Typically, those who take cash advances for non-emergencies are in a credit hole.

If you find yourself in a credit hole, avoid making only a minimum payment. Pay as much of the balance as you can. Do not skip or make late payments; this behavior is reflected on your credit report.

Only take a loan for necessities such as reliable transportation, education, a home, or an emergency. Use the loan funds wisely, do not overspend, and only purchase items you can afford based on your income.

Topic Situation

TOPIC RESPONSE

What are the risks if Simone does not pay the credit card balance in full each month?

Simone has been receiving credit card offers and is deciding whether to apply for one. Simone knows credit cards can be dangerous and can cause financial trouble. However, she also realizes that good credit is needed in order to get a car loan and decides that getting a credit card would not be a bad idea as long as it was not used on frivolous items. Simone reads all the details on each credit card application, including annual fees, minimum payments, and annual percentage rates. After researching the fine print on the credit offers and identifying all hidden fees, she secures a good credit card but uses it only for establishing credit, paying it in full each month.

Credit Reports

When applying for credit, lenders consider your character, capacity, collateral, and condition. Character reflects past behavior toward paying your bills on time, thus communicating to the lender if you will likely repay the loan. Capacity is your ability to repay the loan; your salary will play an important role in this matter. Collateral are the assets you own that are used as security to pay the debt, and conditions are the factors that could potentially harm your ability to repay (e.g., a farmer operating in drought conditions). An important element of securing a loan is the review of your credit report. A **credit report** is a detailed credit history on an individual. Creditors use this report to decide who is a good candidate for credit. A credit report details balances and payments on current and past credit cards and loans. It shows if you have paid these debts on time or if you don't pay them at all. Credit reports are summarized in the form of a credit score. Your credit score is a rating system that evaluates the risk of lending you money based on your credit history. The most common credit rating is known as a **Fair Isaac Corporation (FICO) score**. FICO credit scores have a 300–850 score range. The higher the score, the lower the risk you are to the lender. Therefore, the. higher your FICO score, the better your credit and the better chance of you securing a loan at a lower interest rate. If you have a low FICO score, you have a poor credit rating and you may have difficulty securing a loan and will pay a higher interest rate if you are granted a loan.

The credit report contains personal identification information. This includes any previous names, addresses, and employers. Liens, foreclosures, and bankruptcies will also appear on this report. If you are denied credit because of information on your credit report, the institution is required by law to provide you a copy of your credit report.

There are three credit reporting agencies. They are Equifax, Experian, and TransUnion. Your FICO score is a combination of these agency ratings. Under federal law, you are entitled to a free copy of your credit report from these agencies once every 12 months. For details, please visit the annual credit report authorized by Federal law. While there are many other websites that offer free credit reports, this is the only site that is sponsored by the three national credit-reporting agencies and is affiliated with the national free credit-report program. Take advantage of this free benefit and regularly monitor your credit. Because you can receive one free report from each agency every year, it is recommended you request a free copy from each reporting agency at different times throughout the year. For example, request a copy from Equifax in January, a copy from TransUnion in May, and a copy from Experian in September. Doing so allows you to monitor your credit for free throughout the year. If you find an error on any of these reports, immediately notify the credit-reporting company and correct the error.

Savings and Investments

Do not wait until you have acquired your career job to start a savings plan. Begin saving today. A good rule of thumb is to have five to eight months' income saved for emergencies or major expenses that you did not expect. Save this amount by spending less than you earn. After you complete your budget, you will be able to identify unnecessary expenditures that you can convert into savings.

Keep your savings in a financial institution where it can earn interest. If your company provides an automatic deduction service, use it. An **automatic deduction plan** automatically deducts funds from your paycheck and places them into an account. Make a commitment to take a specific percentage, about 5 percent, from your paycheck and place it in a savings account on a monthly basis.

When saving, decide if you need the money to be readily available or if it can be left untouched for a specified period of time. This will help you determine if you should place your money in a traditional savings account or place it in a certificate of deposit (CD). A traditional savings account typically pays a lower interest rate than a CD. However, you can add and take out funds at any time from a traditional savings account without penalty. A CD pays a higher rate than a traditional savings account, but the funds are locked in for a specified time period. You are not allowed to add funds to the amount during the specified time period. If funds are withdrawn before the maturity date, you will pay a penalty.

There is a difference between saving and investing. Saving money means that you are setting away funds for short-term goals and/or emergencies. Investing may provide a greater opportunity to increase the value of your money and generally is a long-term endeavor. Typical investments include stocks, mutual funds, and real estate. However, investing involves risk. It is recommended that individuals first establish a traditional savings account for emergencies. Once an emergency fund is established, funds can be directed to an investment account.

Investments for the future are important to start now. If invested properly, money grows over time. There are many ways to invest money, all of which you should research to decide what level of risk you desire. Do not invest all of your money or invest it all in one place.

Identity Theft

Protecting yourself from identity theft has become increasingly important. Identity theft is when another individual uses your personal information to obtain credit in your name. To decrease the likelihood of this happening, carefully review your monthly bank and credit statement to ensure charges are valid and accurate. If there are discrepancies, research and report them immediately. Securely store banking statements, credit card statements, and other financial documents for three years. The most popular pieces of personal information desired by identity thieves include your Social Security number, date of birth, credit card numbers, and mother's maiden name. Cut up or shred any communications (electronic or hard copy), including unwanted credit card offers and junk mail that contain your personal information. Register to opt out from receiving credit offers. This can be done by searching "opt out junk mail." Make a copy of your license, Social Security number, and all credit information and store this information in a safe place. Do not share your Social Security number, birthplace, birthday, or mother's maiden name unless you have verified that this individual works for the company from whom you want to secure credit. This private information is used to verify your identity and credit history. If this information gets in the wrong hands, it provides someone easy access to your identity.

The following are tips to remember:

- Do not share your Social Security number over the telephone or Internet without verifying the authenticity of the company and individual requesting the information.
- When using the Internet, note that a secure website will use a Secure Sockets Layer (SSL), which will typically display an icon of a lock by the web address.
- Document all important numbers, such as license, credit cards, and savings account, and keep them in a private and safe place.
- Practice good personal financial management by routinely reviewing details on your credit card bills, bank statements, credit reports, and other financial documents.
- Delete your name from credit card lists and marketing lists.
- Monitor your credit and bank accounts regularly.
- If you receive a call from a collection agency and do not have poor credit, do not ignore the call. Someone may have taken credit in your name and made you a victim of identity theft.

If you become a victim of identity theft, the first thing to do is file a police report. Immediately contact your bank, all credit card companies, and your wireless communications provider. Do not change your Social Security number, but do contact the Social Security Administration Fraud Department and all of the three credit report agency fraud lines. Document everyone you talk to and everything you do for future reference.

Exercise 2.3

Identify ways identity thieves get their information.

_____

_____

_____

Additional Financial Matters

As you've learned throughout this chapter, personal finance is an integral part of your life plan. With spending and saving, consider the long-term impact of your financial choices. As you monitor your finances, attempt to identify when your spending is linked to your emotions. Emotional spending can do significant harm to a budget. Prior to making an unplanned purchase, ask yourself why you are making the purchase and rethink the purchase if it is to fill a void. When out with friends and it comes time to pay the bill, think twice prior to offering to collect the cash and place the balance on your credit card. Chances are you will spend the cash long before the credit bill comes due.

Be cautious when lending large sums of money or co-signing for a loan. When co-signing loans, you assume 100 percent responsibility of that loan. If the other party does not pay the debt, it will result in lowering your credit score. Store all financial documents in one area, preferably in a fireproof safe. If you don't have a safe, make every effort to keep all your financial information secure. Form a long-term relationship with a reputable financial institution such as a bank or credit union. Doing so not only provides you a consistent place for cash and savings transactions, it also creates a resource for other financial services you may need, including check-cashing services, loans, safe deposit boxes, and long-term investment options.

Finally, because an individual's financial matters are personal, make personal finance a consideration when selecting a spouse or life partner. How an individual spends his or her money reflects the individual's values and lifestyle. Pick someone who shares your life goals and financial philosophy. Doing so will increase your communication, trust, and probability of successfully achieving those goals together.

Credit and Fraud Resources

The following is a list of important resources to assist you with credit and fraud issues.

Search online for these credible consumer counseling services:

- The National Foundation for Credit Counseling
- Myvesta Foundation

Search online for these credible credit reporting agencies and resources:

- Equifax Credit Reports: 1-800-685-1111; fraud: 1-800-525-6285
- Experian Credit Reports: 1-800-397-3742; fraud: 1-800-397-3742
- TransUnion Credit Reports: 1-800-916-8800; fraud: 1-800-680-7289
- Free Credit Reports: 1-877-322-8228
- Reporting Social Security Fraud : 1-800-772-1213

MyStudentSuccessLab Please visit **MyStudentSuccessLab**: Anderson|Bolt, Professionalism Skills for Workplace Success, 4/e for additional activities, resources, and outcomes assessments.

Workplace Dos and Don'ts

Do create good financial goals	*Don't* use credit cards unwisely
Do keep a budget	*Don't* waste money
Do start saving and investing now	*Don't* ignore credit reports
Do learn to protect yourself from identity theft	*Don't* use cash for all spending

Concept Review and Application

You are a Successful Student if you:

- Create and implement a personal budget, including savings and investment goals

- Integrate your newfound knowledge of personal finance into your life plan

- Research and explain the optimal types and uses of credit

Summary of Key Concepts

- Personal financial management is the process of controlling your income and expenses.

- Income is money coming in.

- Expense is money going out.

- A budget is a detailed plan for finances.

- The first step to creating a budget is to identify goals.

- Debt (liability) is the money you owe.

- Net worth is assets minus liabilities.

- A credit report is a detailed credit history.

- Identity theft is when another individual uses your personal information.

Self-Quiz MATCHING KEY TERMS

Match the key term to the definition using the identifying number.

Key Terms	Answer	Definitions
Assets		1. Expenses that do not change from month to month
Automatic deduction plan		2. Items that you own that are worth money
Budget		3. The cost of borrowing money
Credit report		4. Small expenditures that consume a larger portion of one's income than expected
Debt		5. A detailed credit history on an individual
Expense		6. Money coming in
Fair Isaac Corporation (FICO) score		7. The amount of money in a paycheck before paying taxes or other deductions
Fixed expenses		8. When funds are automatically deducted from an employee's paycheck and placed into a bank account
Flexible expenses		9. A large debt that is paid in smaller amounts over a period of time and has interest added to the payment
Gross income		10. The amount of money you have after all taxes and deductions are paid
Income		11. Money going out
Interest		12. The process of controlling personal income and expenses
Liability		13. A detailed financial plan used to allocate money for a specific time period
Loan		14. An obligation to pay what you owe
Money wasters		15. The amount of money that is yours after paying off debt
Net income		16. The most common credit rating
Net worth		17. Expenses that change from month to month
Personal financial management		18. Money owed

Think Like a Boss

1. You need to hire a receptionist that will be handling cash. What steps would you take to make sure you hire the right person?

2. Why should you teach your employees the importance of personal financial management? What are creative ways of doing this?

Activities

Activity 2.1

Determine your monthly income by completing the following information the best you can.

Salary/wages per month (use net income—after taxes)	
Interest income per month (savings, checking, other)	
Other income per month	
Total monthly income	

Activity 2.2

If you have debt, create debt-paying goals by listing each loan, the amount you pay each month, and the total amount you owe. Include the amount of interest you pay annually. Identify which creditor should be paid first.

Creditor (e.g., credit card company, retailer, bank)	Amount Paid per Month	Total Amount Still Owed	Interest Percentage	Order of Payoff

Activity 2.3

Using the goals you identified in Activity 1.4, create financial goals to support your life plan. Identify the amount of money needed to reach each goal. In the last column, identify the year you plan to reach each goal.

Life Area	Goal	Financial Goal (Estimated Cost)	Year of Completion
Personal			
Long-term goal			
Short-term goal 1			
Short-term goal 2			
Short-term goal 3			
Education			
Long-term goal			
Short-term goal 1			
Short-term goal 2			

Life Area	Goal	Financial Goal (Estimated Cost)	Year of Completion
Short-term goal 3			
Career			
Long-term goal:			
Short-term goal 1			
Short-term goal 2			
Short-term goal 3			

Activity 2.4

Your instructor will distribute a spending record, or you may create your own. Record all income and spending for two weeks to identify where you are spending your money. Even minor expenses (every penny) should be recorded. Include the date, amount spent, what you bought (service or product), and how you paid for the item.

Date	Amount Spent	Item Bought	Payment Method (cash, credit card, debit card, check)

Activity 2.5

Create a personal budget by only completing the estimated column. After one month, record actual spending and then calculate the difference to identify areas to decrease spending or opportunities to save.

BUDGET FOR MONTH OF _____			
Monthly Payment Category	**Estimated Amount**	**Actual Amount**	**Difference + OR −**
Net spendable income per month	$	$	$
Expenditures			
Clothing			
Communications (phone, Internet)			
Day care			
Debts (including student loans)			
Donations			
Education			
Entertainment			
Food			
Housing			
Insurance			
Investments/savings			
Medical			
Miscellaneous			
Transportation			
Utilities			

3
Time and Stress Management and Organization Skills

resourceful • calm • efficient

After studying these topics, you will benefit by:

- Recognizing how stress affects performance
- Examining the types, causes, and methods of dealing with stress
- Identifying and utilizing time management tools
- Dealing with procrastination to improve personal productivity
- Stating how organization affects time and stress management
- Naming and applying organizational techniques to academic and workplace success

HOW DO YOU RATE?

Is your life in order?	Yes	No
1. The inside of my car is usually clean.	☐	☐
2. My personal workspace is free of clutter.	☐	☐
3. My computer files are in order and it is easy to find documents.	☐	☐
4. I maintain an address book (electronic or traditional) to manage my professional network.	☐	☐
5. I make my bed every day.	☐	☐

▶ If you answered "yes" to three or more of these questions, you are on the path to optimal organization. Organization in all areas of your life decreases stress and improves time management—two factors that will contribute to workplace success.

The Impact of Stress on Performance

Walk into a workplace and you'll quickly form an impression of the work environment. Your first impression will most likely be based on the demeanor of the employees and their interactions with each other. You will also notice if the work area is messy and disorganized or if it is clean and orderly. This chapter examines the influences that stress management, time management, and organization have on workplace productivity. Items arranged in an organized manner make our jobs easier and save us time. When we fail to plan appropriately and do not have enough time to complete our work, we get stressed. Of course, there are other factors that contribute to a productive workplace, but time, stress, and organization are certainly major contributing forces. Stress management, time management, and organizational ability are personal skills that must be developed and consistently practiced. As you learned in Chapter 1, positive personal habits spill into the workplace and become positive workplace behaviors. Employers need employees who are healthy, relaxed, and well organized. Healthy employees are able to perform at their highest levels, have decreased absenteeism, and have fewer health claims than their unhealthy counterparts.

Stress is the body's reaction to tense situations. Stress also affects workplace productivity and is influenced by self-care matters such as diet and exercise and organizational issues like time management. Stress can cause more than just a bad day. Constant stress can result in permanent mental and/or physical harm.

Although some stress keeps you mentally challenged, long-term (chronic) stress will eventually harm you in one way or another. It may start to affect both your work performance and personal life. While not all stress is within your control, try to maintain a low stress level. Stress-related losses are high and, according to the World Health Organization, costs U.S. businesses an estimated $300 billion dollars a year.

Types of Stress

How can stress from school affect other areas of your life?

You arrive in class and your teacher announces that today students are to give impromptu presentations on the lecture material. The students who are prepared and confident may be quite excited about the activity, whereas those who are not prepared or not confident presenting in public may suddenly flush and feel their hearts racing. As a result, they will be stressed. This illustration demonstrates that stressful situations vary from individual to individual. Stress is a normal part of life. What is important is that you recognize when you are stressed and deal with the stress appropriately. You will experience stress at school, at work, and at home. There is no avoiding it. However, how you react to and deal with stress determines how it will affect you. Some stress is minor and affects you at a specific time. This can be **positive stress**. Positive stress is a productive stress that provides strength to accomplish a task. However, even positive stress can become negative if it continues and becomes problematic. For example, if you have a rushed deadline for a special project, your adrenaline will increase, giving you the mental and/or physical strength to finish the project on time. However, if you consistently have rushed deadlines, your stress level can increase and will eventually start working negatively on your mind and body.

Negative stress causes you to become emotional or illogical. This type of stress may affect your mental and/or physical health. Negative stress commonly results in anger, depression, and/or distrust. Other signs of negative stress may include frequent headaches, fatigue, diminished or increased appetite, a poor immune system, or other physical weakness. Continuous negative stress can ultimately result in ulcers, heart disease, or mental disturbances.

Talk It Out

What are common negative stressors students face and possible positive responses to these stressors?

Topic Situation

Dylan has started experiencing headaches and fatigue. After thinking about recent activities, he realizes the headaches and fatigue may be a symptom of stress. With college and a job, there seems to be no time for relaxation. Dylan decides that his situation needs to change or his physical symptoms may get worse. He makes time to reevaluate his goals, write a plan, and identify stress management techniques to help him through this challenging period. Soon after, Dylan feels more control in balancing school and work, and has found free time to relax.

TOPIC RESPONSE

What other symptoms of stress might Dylan experience?

Exercise 3.1

List at least three significant stresses that you have experienced in the last year. Write the result of the stressor, including how you responded mentally and/or physically.

_____

_____

_____

Dealing with Stress

The first step in dealing with stress is to identify key stressors in your life. Learning to both identify and deal with these stressors will reduce their negative effects. Be aware of them and how they affect your attitude and behavior. Life is not stress free. The following steps will assist you in not allowing stressful situations to get the best of you:

1. Find out what is causing you to be stressed.
2. Recognize why and how you are reacting to the stressor.
3. Take steps to better deal with the stress by visualizing and setting a goal for responding in a positive manner.
4. Practice positive stress relief.

Topic Situation

Grace has been noticing that a coworker, Zoey, has been short-tempered and moody lately. Because Zoey is normally very pleasant to work with, Grace decides to ask her if something is wrong. Visiting with Zoey, Grace finds that Zoey is being harassed by someone at work. Zoey tells Grace how stressful this has been and that it is affecting her work and personal life. Grace encourages Zoey to take steps to stop this harassment (presented in Chapter 12). Grace also gives her tips to help deal with the stress. After a few weeks, Grace notices a positive change in Zoey. Dealing with the problem, along with using stress relievers, is helping Zoey get back to her pleasant self.

TOPIC RESPONSE

What advice would you give a friend who has noticeable stress?

Think About It

Although commonly used to relieve stress, what effects do alcohol and drugs have on your body?

Ignoring stress does not make it go away. Being aware of what causes your stress helps you change how it will affect you.

There are common strategies to help relieve stress, including diet and exercise. A healthy body leads to a healthy mind. Consistently eat a balanced diet, including breakfast, lunch, and dinner. At these meals, balance protein, carbohydrates, vegetables, and fruit. Do not skip meals, especially breakfast.

Along with a balanced diet, exercise is essential. When you exercise, your body produces endorphins, which are chemicals that make you feel good. These endorphins help improve your mood, increase sleep, and reduce depression and anxiety. Exercise is also a good way to clear your mind of troubles and increase creativity. You do not have to join a gym or lift weights; you only need a consistent exercise plan that keeps your body moving. There are simple ways to increase physical activity, including using the stairs instead of taking the elevator or parking your car a little farther away from a building to increase your walking distance. Exercising for 10 minutes several times a day will increase energy and improve your health. Diminish—or ideally, eliminate—the use of alcohol and/or drugs. These stimulants may cause mood swings that typically make matters worse. Though common among college students, lack of sleep is also a contributor to stress. Sleep deprivation contributes to obesity, depression, and other chronic diseases. The Center for Disease Control recommends adults receive 7–9 hours of sleep per night. If you are not consistently waking up refreshed, you most likely are sleep deprived. Take small steps to gradually change your sleep pattern. Begin going to bed earlier; limit caffeine intake prior to bedtime; and sleep in a quiet, dark space (even if it includes ear plugs and a sleep mask). When you are able to consistently wake up on time, without an alarm clock, you body is most likely getting the amount of sleep it needs. When your body gets enough sleep, you will see a noticable improvement in your energy level, your attitude, and your productivity.

Web Search

Find an app or website with nutritional resources to help you maintain a healthy diet.

There are other simple physical activities that relieve stress that you probably do without realizing their benefits. These include enjoying leisure time, listening to music, meditating, deep breathing exercises, and using positive visualization.

Recognizing what situations cause stress allows you to better control them. The more organized you are, the better prepared you will be, thus reducing stress.

Keep your emotions in check. Becoming emotional means you are losing control and risk becoming illogical in your response to the stress. When at work, if you cannot surround yourself with positive people, create a positive personal space where you can take a few minutes for yourself. Realize that people are not always going to agree with you at the workplace. There may be annoying people, and there may be people with whom you may not have a positive relationship.

You may find yourself in situations that become very stressful. Use the stress relief methods mentioned earlier in this chapter and make the best of the situation. As we discussed in Chapter 1, only you can control your attitude and your response to challenging situations.

Take time outside of work to relax. Do not bring work troubles home with you, nor take home troubles to work. When you recognize personal stressors and take care of yourself, you can reduce and/or eliminate the impact stress may have both at home and at work. Create and maintain a support network. Identify a few close friends and family members in whom you can confide and share concerns. Develop realistic goals.

If your company offers an employee assistance program (EAP), use it to get professional help. Typical employee assistance programs offer help with financial, legal, and psychological issues. Additional information on employee assistance programs is provided in Chapter 12.

Job burnout is a form of extreme stress where you lack motivation and no longer have the desire to work. Factors that lead to job burnout include not being able to control decisions that affect your job; being unclear of your job duties; working with bullies, negative colleagues, or a bad boss; and not enjoying your job or career.

Signs of job burnout include:

- Frequent tardiness or absenteeism
- Continually complaining or gossiping
- Exhibiting poor physical and emotional health
- Lacking concern for quality
- Clock-watching and being easily distracted
- Lack of satisfaction in your work
- Demonstrating a desire to cause harm to the company (theft of or damage to property)

Determine the source of job burnout and take steps to deal with or eliminate the issue before it causes significant damage. If you have seriously tried to improve the current work situation and still find yourself at a dead end, you may need to consider a job change. Continuing in a job in which you have not been motivated in for a long period of time is destructive not only to you, but also to your company and coworkers.

Web Search

Conduct a web search for an online quiz that will help you measure your stress level.

Exercise 3.2

What can you do if you begin feeling job burnout?

_____

_____

_____

Time Management

Recall the earlier scenario in which the teacher assigned impromptu presentations on the lecture material. Perhaps some students were stressed because they did not study and therefore were not prepared. There is a clear link between stress and time management. There is also a link between time management and success. **Time management** is how you manage your time. In business, time is money. The ability to use time wisely is a skill in itself—one that is necessary in the workplace. When you use your time efficiently, your tasks will be completed on time or even early. Without proper time management skills, you may forget, lose, or spend more time than needed on an important project. Proper time management at work frees up more time for other activities, both at work and at home. If you are being efficient and paying attention, your employer sees that you care about your job and are organized. In turn, this may lead to higher pay and/or a promotion.

You may get stressed at work because you do not have enough time to complete a project. However, many work projects are similar in nature and can therefore be managed easily. Prior to starting a project, make a plan. Set priorities and get organized. Do not wait until the last minute. If you have similar projects, create a template so you are not starting over with each project. Focus on completing a job right the first time; rushing through a job typically results in errors that will only take more time to correct.

A common workplace interruption is that of individuals who visit your work area and stay longer than necessary. When dealing with these individuals, always be professional and polite. Inform the individual that although you would like to visit, you have work that must be completed. If you are in an office environment, do not sit down or invite your office visitor to sit. Standing by the door or entry to your workspace, politely tell your visitor that you are busy and unable to visit. Avoid having items on your desk that attract unwanted guests such as a candy dish.

Break larger tasks into simpler, smaller ones. When you break down tasks, you can space out projects. This enables you to organize the time needed to complete each task before starting the next. Again, the exception to this rule is if you have a priority task that needs to be completed immediately.

Further discussion of the use of body language and communication will be presented in later chapters.

At the end of the chapter, Activity 3.1 provides the opportunity to identify how you are currently spending your time.

The following tips will help you organize and control your time:

1. Make a list of tasks for each day and prioritize that list; this is commonly referred to as a *to-do list*. Many PCs and smartphones offer task applications to make electronic lists.
2. Keep a calendar accessible at all times. List all appointments, meetings, and tasks on your personal electronic or traditional calendar.
3. Organize your work area. Use file folders and in-boxes to organize and prioritize projects, including your computer desktop and files.
4. Practice a one-touch policy. After you have looked at a project, letter, memo, or other item, either file it, place it in a priority folder, forward it to the appropriate individual, or throw it away. Do not pile papers on your desk.
5. Answer memos that only require a short response by writing the response directly on the original memo and keeping a copy for your records.
6. Avoid time wasters. Time wasters are small activities that take up only a small amount of time but are done more frequently than you may realize. These include unnecessary visiting or inappropriate activities such as personal texting or participating in social networking.
7. If possible, set aside time each day to address all communication at once during a certain time of the day, as opposed to handling messages as they arrive (e.g., e-mail and phone messages).
8. If needed, ask for help. Asking for help is not a sign of weakness or inefficiency if you are practicing sound time management techniques.

Exercise 3.3

List time wasters you have experienced in the past few weeks. How do these time wasters affect productivity? What change should be made?

_____

_____

_____

Procrastination is putting off tasks until a later time. This poor habit severely impedes time management and contributes to stress. People procrastinate for many reasons, including fear of failure, perfectionism, disorganization, or simply not wanting to perform the task because it is not pleasant. Procrastination can lead to the loss of opportunities. As a result of a late project, you may lose money, lose respect from coworkers and/or your boss, or not be as successful as possible. To overcome procrastination, first visualize the completed task. Knowing your end result and how you will feel when it is completed will motivate you to get started. The next step is to make a plan for completion by identifying what information and resources are required for the end result you envisioned. List every activity and piece of information you will need. After you have made your plan, get to work. If the task appears overwhelming, break it down into smaller tasks and complete each task in priority order. Breaks and celebrations are not only essential, but encouraged when you are working on a big project.

Copyright © 2016, 2013, 2011 by Pearson Education, Inc.

Think About It

List a recent time when you procrastinated on a project. What was the reason for procrastination? What was the result?

Topic Situation

Jonelle was taking a chemistry class in which the instructor assigned a semester-long research project. When the project was assigned at the start of the semester, the instructor encouraged the students to make a plan and schedule dates to complete sections of the research throughout the semester so as to produce a quality project. Jonelle struggled with the class material and procrastinated working on the assignment. Unfortunately, as the semester wore on, Jonelle became immersed with other courses, a job, and personal issues and kept delaying the research project. The more Jonelle thought about the project, the more stressed she became. Finally, two weeks before the end of the semester, the instructor reminded the class that all research projects were due the day before the final. Jonelle realized there was no time to properly study for exams and also complete the research project. The procrastination resulted in her being stressed and receiving a failing grade in the chemistry course because she gave up and did not even attempt to write the paper or take the final exam.

TOPIC RESPONSE

What steps could Jonelle have taken to avoid procrastination in this situation?

One final issue that contributes to both stress and poor time management is the inability for individuals to say no to coworkers, bosses, or others. At work, our goal is to be as productive as possible by prioritizing our current workload. Overcommitting ourselves risks compromising quality for quantity.

When you are pressed for time and someone asks you to assist with a project, first evaluate if the project is part of your primary work duties. If it is not your job and it does not conflict with your priority projects, agree to take on the new project if you have the time. If you do not have time and you have greater priorities, decline the project. If it is your boss that is making the request, politely inform your boss that you want to help wherever and whenever possible, but you are currently working on another priority project and ask him or her which project should take precedent. Many bosses are unaware of an individual's workload at any given time, so your goal is to communicate your current priorities.

Talk It Out

What other school-related activities do students procrastinate getting completed?

Organizing and Performance

Individuals who are organized operate around goals and have learned that being surrounded by clutter deters focus. Organized individuals arrange their belongings in their homes and work environments in a manner that reflects their goals.

In Chapter 1 you established a life plan and created goals to support this plan. Your life plan details what you want to accomplish and by when. Getting organized for optimal performance is not difficult and will result in your using your time more efficiently and reducing stress.

Although it may take time to organize, the time you invest in cleaning and organizing your space will release much more time for you to accomplish your goals. An organized and clutter-free area is calming and allows you to focus.

Tools for getting organized in the workplace can also be used at home. Technology has made it easier to get organized with electronic devices. However, there are other common organization tools to use, including shredders, files, and desk space organizers.

One of the easiest ways to get organized is to use a calendar. There are many options, including a computerized calendar, a mobile calendar, and a traditional paper calendar. For efficiency, businesses prefer an electronic calendar for computer networking purposes. It is common to have access to a computerized information manager on the web, a computer, and/or a mobile device. Determine which type of calendar works best for your work situation; sometimes the solution is to use more than one calendar and sync them. Once you have determined which option is best for you, make a commitment to record all work-related and personal meetings and important deadlines. If your personal information manager and communications program is electronic, store telephone numbers, e-mail addresses, and other important messaging data in the program for easy access. Tasks, to-do lists, and notes can also be monitored and updated. Keep data current by immediately recording changes. If you use multiple organization tools, make a habit of transferring information on a daily basis, or connect with web programs that automatically sync and update all your electronic devices. For maximum efficiency, customize applications to suit your needs.

Other ways to keep organized and improve performance is to check and answer your phone messages and e-mails at specific intervals. It is inefficient to return each phone message or e-mail as it comes in. The only exception is when there is an important message or e-mail that needs to be sent or answered immediately.

If you are assigned a personal workspace, keep your work environment and desk clean and clutter free. Maintain a professional look by having only a minimal number of personal items on your desk. Take inventory of your workspace; if you don't need an item, remove it from your work area. If an item is necessary but not used often, store it. Keep frequently used work tools easily accessible, including a stapler, tape, a notepad, pens, pencils, paperclips, scissors, a ruler, a calculator, highlighters, and a computer storage device. Place items where they are required (i.e., printer paper by the printer, notepad near the phone). Return items to their appropriate area after use. In addition, the use of a small bulletin/whiteboard for posting important reminders will help you keep track of important tasks and appointments. Have a trash can close to your desk, and throw away supplies that have been used or do not work anymore. Shred confidential materials at least once a day.

When managing paper files, maintain these files properly in a file cabinet and keep files neatly arranged in clearly labeled file folders. Avoid miscellaneous piles and folders. File dated documents in chronological order (most recent first). Other files can be arranged by subject or alphabetically. Be consistent in your filing method. Routinely used files should be easily accessible. Files should be updated and old files disposed of properly. Any unnecessary files with personal information or identification numbers are considered confidential and should be shredded. If files are not important and do not have identification, they may be thrown in the trash.

For efficiency and security purposes, keep electronic files organized. Your computer desktop should contain only shortcuts to frequently used programs and files that you are currently working with. Routinely clean your computer desktop to ensure it is clutter free. Just as with paper files, electronic files should be well organized and labeled. Establish folders for major projects, committees, and other items related to your job. Place appropriate documents inside the respective major project folder. Whenever possible, create subfolders for large projects so that you can properly file and quickly retrieve documents when necessary. Keep both folder and file names simple and easily identifiable. Also remember to routinely back up and/or secure your files to protect confidential information.

Effective organization includes the proper handling of both electronic and paper mail. Your job may include sorting and/or opening mail. Use a letter opener to open all paper mail at one time. After opening the paper mail, sort it into piles. Throw away or shred junk mail immediately after opening. Respond to the sender of the mail if needed, file the document, or forward the mail to the appropriate party within the company. Do not open mail that is marked confidential unless instructed to do so. Mail should be kept private and not shared with coworkers. If you encounter a piece of mail that should be confidential, place it in a separate envelope and mark it confidential. Company letterhead or postage is not for personal mail.

MyStudentSuccessLab | Please visit **MyStudentSuccessLab:** Anderson|Bolt, Professionalism Skills for Workplace Success, 4/e for additional activities, resources, and outcomes assessments.

Workplace Dos and Don'ts

Do recognize your stressors	*Don't* let stress go until you get mentally or physically sick
Do deal with stress appropriately	*Don't* think that stress will just go away
Do eat a balanced diet and have an exercise plan	*Don't* skip breakfast
Do manage your time by setting priorities	*Don't* be afraid of asking for help when getting behind
Do take time to get organized	*Don't* give in to time wasters

Concept Review and Application

You are a Successful Student if you:

- Apply the tools in this chapter to create a stress management plan
- Create and utilize a calendar system
- Summarize ideas for organizing your work area

Summary of Key Concepts

- Stress is a physical, chemical, or emotional factor that causes bodily or mental tension.
- Stress can be positive or negative.
- Signs of stress include becoming emotional or illogical or losing control of your temper.
- The first step in dealing with stress is to identify the stressor.
- A balanced diet along with exercise will help you to better manage stress.
- There are many ways to reduce stress, such as setting goals, relaxing, and getting enough sleep.
- Good time management comes from being organized.
- Avoid procrastination.
- Being organized will optimize your performance and reduce stress.

Self-Quiz   MATCHING KEY TERMS

Match the key term to the definition using the identifying number.

Key Terms	Answer	Definitions
Job burnout		1. Putting off tasks until a later time
Negative stress		2. How you manage your time
Positive stress		3. A body's reaction to tense situations
Procrastination		4. A form of extreme stress where you lack motivation and no longer have the desire to work
Stress		5. Productive stress that provides strength to accomplish a task
Time management		6. Unproductive stress that affects your mental and/or physical health

Think Like a Boss

1. You have noticed that an employee is frequently calling in sick and appears agitated when at work. What do you do?

2. You have just become the supervisor for a new department. What can you do to make the department and its employees more organized? Discuss appointment tools, necessary equipment, and software.

Activities

Activity 3.1

Your instructor will distribute a time log or you may create your own. For the next 24 hours, use this log to track how you spend your time. Account for every minute. When you are finished, identify specific time wasters.

List three time wasters from your time log
1.
2.
3.

Activity 3.2

In addition to those mentioned in this chapter, research physical responses generated by prolonged stress. List your findings.

1.		4.
2.		5.
3.		6.

Activity 3.3

List five time management tools commonly used in your target career.

Tools
1.
2.
3.
4.
5.

Activity 3.4

A nutritious diet can make a difference in how you perform throughout the day and how you react to stressful situations. List what you have eaten in the last 24 hours, recording the time of day, your mood or situation, and if the item was nutritious. Evaluate if changes need to be made to your diet.

Time of Day	Food	Mood/Situation	Was It Nutritious?

Activity 3.5

Identify one space at home you need to organize. Create a plan to overcome your procrastination of dealing with this issue.

Space	Plan

4

Etiquette/Dress

impression • manners • perception

After studying these topics, you will benefit by:

- Explaining the elements of professional dress
- Recognizing the importance of making a positive first impression
- Expressing an understanding of workplace etiquette
- Identifying the importance of making and keeping appointments
- Describing the impact dress can have on others' perception of you
- Demonstrating appropriate behavior in work-related social situations

HOW DO YOU RATE?

How proper are you?	True	False
1. You do not have to shake someone's hand if you already know the person.	☐	☐
2. Visible tattoos, nose rings, or lip rings, if tasteful, are now acceptable in a professional business situation.	☐	☐
3. If you are invited to a business meal, you may order anything on the menu.	☐	☐
4. Sending a thank-you note is no longer necessary.	☐	☐
5. It is now acceptable business practice to read text messages during business meetings.	☐	☐

▶ If you answered "false" to four or more of these questions, you are actively practicing business etiquette. Though business protocol may vary in some industries, it is best to lean toward a conservative, traditional approach until you are confident of acceptable industry standards.

Employees represent their company. Therefore, the way you communicate, dress, and behave both inside and outside the company contribute to others' perception of you and your company. Consistently demonstrating proper etiquette and protocol in business, dining, and social situations results in positive business relationships. The way you look and behave is a reflection of the organization for which you work. The purpose of this chapter is to prepare you for many of the social experiences you will face in the workplace.

Executive Presence

Think About It

Do you practice good manners on a regular basis?

Executive presence is defined as having the attitude of an executive by demonstrating appropriate workplace behavior. Projecting an executive presence demonstrates to employers that they have hired a new employee with knowledge regarding appropriate workplace behavior.

Many of us have parents who taught us early in life that good manners, such as smiling and saying please and thank you in social situations, create positive relationships. To be successful at work, manners are important, along with understanding the basics of professional attire, business protocol, social etiquette, dining, and the appropriate use of technology. You will encounter business-related social situations at work, and knowing how to behave will help you be more successful in workplace relationships. Some of this information may be new to you, and you may feel awkward when you first implement these positive behaviors.

Influences of Appearance

Both your maturity and the importance you place on your job are reflected in the way you dress and behave at work. Because impressions are often made in the first few seconds of meeting someone, individuals rarely have time to even speak before an impression is formed. The majority of first impressions are made through appearance. Coworkers, bosses, and customers form attitudes based on appearance. Appearance also has an impact on how you perform at work. If you dress professionally, you are more apt to act in a professional manner. The more casual you dress, the more casual you tend to behave. Think of your appearance as a frame. A frame is used to highlight a picture. You do not want the frame to be too fancy, because it will take away from the picture. You want a frame to complement the picture. The frame highlights not only your physical features, but also your attitude, knowledge, and potential.

Exercise 4.1

Define your frame. Be honest. Is it trendy, outdated, professional, or inconsistent? Does it complement your personal brand and desired appearance as a professional? If your current frame is not yet professional, what changes need to occur?

_____

_____

_____

One of the toughest transitions to make when entering the workplace is choosing appropriate dress. Dressing professionally does not have to conflict with current fashion trends. The trick is to know what is acceptable. A basic rule of thumb is to dress one position higher than your current position (i.e., dress like your boss). Doing so communicates that you are serious about your career and how you represent the company. Dressing professionally will assist you in projecting a favorable image at work and position you for job advancement.

Know your workplace dress policies, and understand that professional dress carries different meaning depending on both the industry and work environment. One of the first steps to determining appropriate attire for work is to identify your company's **dress code**. A dress code is a policy that addresses issues such as required attire, uniforms, hairstyle, undergarments, jewelry, and shoes. Dress codes vary by company, industry, the specific work area, and health/safety issues. If your company has a mandatory uniform, the company dress code will be detailed. If a uniform is not required, identify what is and is not acceptable attire by reading the dress code policy, by observing what is practiced in the workplace, or by asking your supervisor. Some dress codes are vague, whereas others are specific. Work attire should pose no safety hazards—for example, unstable footwear that does not provide protection or dangling jewelry that could be caught in equipment are not appropriate. As previously stated, organizational dress policies exist for customer service, safety, and security reasons. Frequently, these policies are included in the employee handbook. If there is no policy, ask your boss if there is a formal dress code and secure a copy. An important cue to workplace attire is to observe how managers dress. Formal business suits are not always the preferred attire in an office environment. Note that sweats (shirts and/or pants) are not appropriate for the traditional workplace.

Once you have identified what your organization considers proper attire, begin to create a **work wardrobe**. These are clothes that you primarily wear to work and work-related functions. You need not invest a lot of money when building a work wardrobe. Start with basic pieces and think conservative when working in a traditional office environment. For women working in a traditional office environment, this attire includes a simple, solid skirt or pantsuit in a dark color and a blazer. Skirt length should not be above the knee. Pants should be worn with a matching blazer. For most office environments, men should select dark slacks, a matching jacket, and a tie. If you are just starting your job and cannot afford new clothing, these items can sometimes be found inexpensively at thrift and discount stores. If these items are purchased at a thrift store, ensure they are not torn, stained, or faded and have them cleaned and pressed. You will be surprised at how professional these items look after they are cleaned and pressed. Select items that are made of quality fabrics that fit properly, are comfortable, and will not wear out quickly. As your income grows, continue building your wardrobe and develop a style that conforms to both company policy and your taste.

Talk It Out

Identify local stores where you can purchase professional attire at a low cost.

Casual Workdays and Special Events

Many companies allow **casual workdays**. These are days when companies relax their dress code. Unfortunately, some employees attempt to stretch the term *casual*. If your company has a casual workday, remember that you are still at work and should dress appropriately. You may wear jeans if allowed; just adhere to the head-to-toe tips presented later in this chapter. Do not wear clothing that is tattered, stained, or torn (even if it is considered stylish). Avoid wearing shirts with sayings or graphics that may offend others. In general, it is best to dress modestly.

As you learn more about professional dress and expectations regarding professional attire, consider cultural and geographic differences and expectations. Globally, differing cultural expectations apply to workplace dress. In some countries, women must be completely covered from head to toe, whereas in other countries, women should not wear pants. Business casual for men on the East Coast of the United States is more conservative and may require a blazer, while on the West Coast a polo shirt is acceptable. When conducting business in a geographic area different than yours (whether in your own country or abroad), research appropriate attire prior to your visit.

Your company may host or invite you to attend a special function such as a holiday party or reception. In these situations, instead of daily work attire, more formal attire may be required. Just as with casual workdays, stick with the basics provided in the head-to-toe tips. Women, if appropriate, should wear something in a more formal fabric. Although you have increased freedom and flexibility regarding style and length, this is still a work-related function, so dress conservatively and not suggestively. Men should check ahead of time to determine if tuxedos are preferred. For most semiformal occasions, a suit will suffice.

Talk It Out

Identify people in class who are wearing something appropriate for a causal workday.

Tips from Head to Toe

Regardless of the company's dress code, men and women should practice these basic dress and hygiene rules:

Hygiene	• Shower daily
	• Use deodorant
	• Use perfume, lotion, or cologne sparingly
	• Avoid overpowering scents
Clothes	• Should fit properly and be clean and ironed, not torn or tattered
	• Sweats (pants or shirts) are not appropriate for work
Hair	• Clean, well-kept, and a natural color
	• Hairstyle should reflect profession
	• Fad hairstyles and unnatural color are inappropriate in many workplaces
Dental hygiene	• Brush and floss teeth both in the morning and at bedtime to ensure clean teeth, fresh breath, and to help prevent tooth decay
	• Schedule routine dental check-ups and cleaning; many public health clinics provide no-cost or low-cost dental care

Hands and nails	• Keep nails clean and well groomed • Unnaturally long nails are inappropriate • Polish or artwork, if allowed, should be neat and conservative
Jewelry	• Keep jewelry to a minimum • Jewelry should complement your outfit • Do not wear anything that is distracting or makes noise • Avoid large and gaudy items
Shoes	• Shoes should be in good condition • Keep shoes polished and free of scuffs • Flip-flops are not appropriate for the workplace
Miscellaneous	• Hoodies are inappropriate • Do not wear sunglasses inside (on face, on top of head, or on back of head) • Ear buds should not be visible

A woman's outfit should reflect her style and personality—within reason. When dressing for work, your goal is to appropriately frame yourself in a manner that draws attention to your professional qualities. Additional tips for women include the following:

Talk It Out

Discuss today's fashions and trends that would or would not be appropriate for the workplace.

Shoes	• Keep heels in good condition; repair or replace them as needed • Heels should not be too high • Though dress sandals and open toed shoes may be appropriate, flip-flops, in any style, are not appropriate
Nylons, tights, and socks	• Nylons should be free of runs and snags • Socks should be a solid color and match outfit
Makeup	• Makeup should be minimal for day wear • Avoid colors or styles that are unnatural or distracting • Makeup is appropriate for work; however, makeup that makes people think you are going to a bar after work is not • Do not wear heavy eyeliner or eye shadow in colors that draw attention • Wear natural color lipstick, no bold colors
Clothing	• No suggestive clothing • Visible cleavage or bare midriffs are inappropriate • Undergarments (bras and panties) must be worn and should not be visible • Skirts should be no shorter than knee length

Just like a woman's outfit, a man's work wardrobe should reflect his style and personality. For some positions, a suit may not be appropriate. The biggest wardrobe blunder men make is wearing clothing that is not clean and/or pressed. After reviewing your company's dress code, heed these unspoken rules regarding professional dress for men:

Shoes and socks	• Shoes should be polished and scuff free • Shoes should match pant color • Socks should match shoe or pant color • No flip-flops or sandals
Facial hair	• Shave and/or trim facial hair, including nose and ear hair
Pants	• Dress pants are the only pants that are professional • Jeans are only appropriate on casual workdays, when allowed • Baggy pants that reveal underwear are inappropriate • Whenever possible, wear a neutral, plain belt
Shirts	• Shirt should be tucked in • Wear dress shirt with a tie • A polo shirt is most appropriate for casual work days • Shirts should not display excessive wear (check around collar line for fraying or stains) • Inappropriate logos or offensive phrases are unacceptable at work
Hats	• Hats should not be worn inside buildings except for religious purposes

Jewelry, Body Piercing, and Tattoos

Think About It

How may a tattoo or piercing affect securing a job in your target career?

As with professional attire, you do not want to wear or display anything that brings unwanted attention to you in the workplace. Although body art and piercings are becoming more common and acceptable in society, many companies have policies that prohibit visible tattoos and/or visible body piercings beyond one in each ear. Body art and piercings are offensive to some individuals. Many people get a tattoo and/or body piercing to signify a special event, individual, or symbol. If you are considering a tattoo or body piercing, think about the long-term consequences of doing so. Relationships and situations change. Consider size, color, graphic, and placement and how it will affect your appearance in professional situations. Tattoos and some piercings are difficult and painful, if not impossible, to remove. Though you may currently not care how society feels about your tattoo and/or piercing, you may regret your decision in the future. If you already have body art and/or piercings, it is recommended that you cover your tattoo with clothing, makeup, or other methods until you are clear on your employer's policy regarding visible body art. Many companies also have strict policies on body piercings beyond earrings.

Some piercings close quickly, so it may be impossible to remove the piercing during work hours. Other forms of piercings, such as microdermal piercings, cannot be easily removed. In these cases, determine which is more important—a job, or your body art and/or piercing. In general, follow these guidelines regarding jewelry, piercings, and tattoos:

- Nose rings, lip rings, facial piercings, tongue rings, and/or plugs are not professional and should not be worn in a professional setting.
- Piercings and/or jewelry on other parts of the body should not be visible at work.
- More than two earrings worn on each ear is considered unprofessional.
- Earrings, chains, and other jewelry should not draw attention. This includes symbols or words that could be considered offensive to others.
- Body art (tattoos) should not be visible at work.

Business Etiquette

In a modern workplace, human interaction is unavoidable. Our society has a standard of social behavior that is called **etiquette**. Typically, when individuals think of etiquette, they think it applies only to high society. This is not true. Socially acceptable behavior should penetrate all demographic and economic groups. Individuals wanting to succeed in the workplace need to heed this protocol and consistently utilize proper etiquette not only at work, but in all areas of their life.

Before we study common areas of business etiquette, we need to define a few terms. Understanding these terms and integrating them into your daily routine will make it much easier to carry out the desired and appropriate workplace behavior. The first word is **courtesy**. When you display courtesy, you are exercising manners, consideration, and respect toward others. **Respect** is defined as holding someone in high regard. This means putting others' needs before your own. Displaying both courtesy and respect toward others is the key to becoming ladies and gentlemen at work. One of the easiest ways to show respect is to treat others as you want them to treat you. Additional ways to show respect include being polite, listening, watching your manners, and showing concern for others.

As mentioned at the start of this chapter, some of the first words most parents teach young children are *please* and *thank you*. Although they are not used as frequently as they should be, both are extremely valuable terms that can actually create power for you at work. Think about it; when someone says "please" and "thank you" to you, you are more likely to repeat a favor or gesture because your deed was acknowledged. When someone does something nice, verbally say "thank you." Not doing so makes you appear selfish and unappreciative. When you express thanks, individuals will be more likely to continue performing kind acts for you.

Make it a habit to write a thank-you note when someone does something for you. An e-mail is acceptable for a simple gesture. However, if the activity took more than five minutes or someone gave you a gift, the thank-you note should be handwritten. Write the note as soon as possible. Do not wait more than three days to write and deliver the thank-you note.

Talk It Out

Discuss ways to demonstrate courtesy in class.

Exercise 4.2

List ways you can earn respect.

_____

_____

_____

Handshakes

A good handshake conveys confidence. Make a habit of greeting others in business situations with a professional handshake and friendly verbal greeting. Approach the individual you are greeting, make eye contact, smile, and extend your right hand as you verbalize a greeting. For example, "Hello Ms. Cao, my name is Talia. We met at last week's meeting. It's nice to see you again." Ms. Cao will extend her right hand. Your two hands should meet at the web (see Figure 4.1). Grip the other person's hand and gently squeeze and shake hands. When shaking hands:

- Do not squeeze the other hand too firmly.
- Shake the entire hand and not just the other person's fingers. Shaking only the fingers is insulting and implies that you feel you are better than the other person.
- Do not place your hand on top of the other person's hand or pat the hand. Doing so implies superiority.
- If your palms are sweaty, discreetly wipe your palm on the side of your hip prior to shaking.

A good handshake takes practice. As mentioned earlier, get into the habit of being the first to greet and introduce yourself to others. At first you may not feel comfortable, but practice makes perfect. The more frequently you initiate a good handshake, the more comfortable and confident you will become.

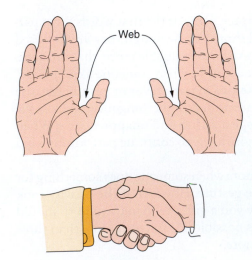

Figure 4.1

Proper Handshake

Exercise 4.3

With another person, practice introducing yourself with a professional handshake. Have the other person rate the quality of the handshake on a scale of 1 to 5, with 5 being the best. Discuss what improvements should be made.

_____

_____

_____

Introductions and Business Networking

An element of success in the workplace involves meeting new people. This process of meeting and developing relationships with individuals outside of one's immediate work area is referred to as **networking**. Networking is also commonly used in the job search process and will be discussed in greater depth in Chapter 13. In the workplace, creating a professional network is a useful tool for collaboration. Each time you meet a new person, treat him or her as you would the president of the company. When you do not know someone in the room, increase your confidence by being the first to initiate a conversation. After you have introduced yourself, ask your new acquaintance about him- or herself. Learn about his or her job and find something you have in common. Keep the initial conversation focused on the other person. Your goal is to meet new people and create a positive impression so that if you see them again or contact them in the future, they will remember and have a favorable impression of you. A positive first impression is created by smiling, making eye contact, and standing confidently.

At times, you will be with individuals who do not know each other. When you are with two people who do not know each other and you know both people, it is your responsibility to introduce them. Politely introduce the lower-ranking person to the higher-ranking person. For example, "Ben, this is Rafaela McClaine, the president of our company. Rafaela, this is Ben Yu, my next-door neighbor." Apply this introduction rule to all social situations, including dining, meetings, receptions, and parties. Making introductions to others is an excellent form of networking. After you have introduced the two individuals, if possible, provide a piece of information about one of the individuals that creates a foundation for a conversation. For example, "Rafaela, you and Ben attended the same college."

Talk It Out

What prevents you from initiating a conversation with someone you do not know? What steps can you take to overcome these barriers?

Exercise 4.4

In groups of three, assume one person to be your teacher, and the other a friend. Practice an introduction of the teacher to your friend.

_____

_____

_____

Appointments

A daily function of business is making and keeping appointments. Appointments can occur in many forms, such as face-to-face meetings, over the phone, or through Internet technologies (e-mail, texting, or video chat). When setting meeting times, check regional time differences and clearly include the regional time zone abbreviation in your confirmation. For example, "I look forward to meeting with you on Tuesday, April 21, at 9 A.M. Pacific Standard Time (PST)."

Sometimes you will be required to work with receptionists and/or administrative assistants to schedule appointments. Be kind to the receptionist and/or administrative assistant. These individuals are the gatekeepers to their bosses; they control schedules and often wield great power in decisions. When scheduling an appointment, state your name, the purpose, and the desired date and time of the meeting. Avoid scheduling appointments on Monday mornings; many people use Monday mornings to organize their work week. If you will be arriving late to an appointment, call and inform the other party that you are running late. There are also risks of arriving too early. Do not arrive more than 10 minutes early to your meeting location. Arriving early may imply that you do not follow directions. When canceling an appointment, do so immediately and apologize for any inconvenience. Do not simply ignore an appointment.

Ensure that meetings taking place over the telephone are held in a quiet location where there are no distracting background noises. Use a reliable phone connection. If a meeting requires Internet technologies, log in at least 20 minutes early to ensure both a proper and reliable connection. When a meeting involves video chat with a web camera, dress professionally and hold the meeting in an appropriate location, such as an office or study. Due to confidentiality issues, problems with noise, and the need for a professional backdrop, do not use a public location. Additional information regarding communications technology is included in Chapter 10.

When keeping an appointment (face-to-face or via technology), arrive or check in five to 10 minutes early, but no earlier. For face-to-face meetings, after you enter the office, greet the receptionist and politely introduce yourself. State whom you have an appointment with and the time of the meeting. When entering an office for a meeting, wait to be invited to sit down. At the close of any meeting, thank the other participants for their time. If you are in person, exchange business cards, if appropriate, and close with a final handshake. Additional information on meeting management will be presented in Chapter 11.

Dining

There are a variety of dining situations that occur in the workplace. Although some dining experiences are less formal than others, you will most likely come across some form of the table setting illustrated in Figure 4.2. Take time to study and review a common table setting to learn the proper location and use for utensils, plates, and cups. Apart from fast food, few college students are generally comfortable eating in a formal dining situation. Here are several rules of thumb regarding dining etiquette:

- As soon are you are seated, place your napkin on your lap. If you leave the table, place your napkin to the side of your plate, not on your chair.
- Do not order anything expensive or messy.

1 napkin
2 plate
3 salad fork
4 dinner fork
5 dinner knife
6 teaspoon
7 soup spoon
8 salad plate
9 bread plate
10 butter knife
11 dessert spoon
12 dessert fork
13 water glass
14 beverage/wine glass
15 coffee cup and saucer

Figure 4.2

Table Setting

- Do not order alcohol unless others at your table first order an alcoholic beverage. Abstaining from alcohol is the most desired behavior. If you choose to drink, limit consumption to one drink.
- Do not discuss business matters until everyone has ordered. Table conversation should be positive and free of controversial subjects such as politics and religion.
- When serving beverages or bread available at the table, first offer and serve others at your table.
- Utensils are set to be used in order of necessity. As your courses are served, start with the outside utensil and work in, toward the plate. The utensils set at the top of the plate are for your dessert.
- Place a serving of bread and a serving of butter on the bread plate. Tear a piece of bread, and butter only that piece of bread before eating.
- Offer the last piece of bread or appetizer to others before taking.
- Begin eating only when everyone at your table has been served. If everyone receives their meals except you, give others at your table permission to begin eating without you. Eat your meal at the same pace as others at the table.
- Do not eat your meal with your fingers unless your main course is intended to be eaten without utensils.
- Be kind and polite to the staff and servers.
- Chew with your mouth closed and do not talk with food in your mouth.
- Burping and slurping are inappropriate. If you accidentally burp or slurp, immediately apologize and say "excuse me."
- When you are finished eating, place your knife and fork together, with the blade facing in and the tines up. When you are only resting and you do not want the server to take your plate away, cross your utensils with the tines facing down.
- It is inappropriate to use a communication device while dining. If you must take a call or text, excuse yourself from the table before talking.
- Pay your portion of the bill and properly tip the wait staff. If you are an invited guest, offer to pay your portion. If the host pays the entire bill, be certain to thank the host when the bill is paid.

Talk It Out

Share common dining and social situations that make you uncomfortable and identify how best to deal with these situations.

TOPIC RESPONSE

What steps would you take to ensure you dress and act appropriately when attending a business conference?

When Briggs arrived at the technology conference, he noticed that everyone was dressed in business attire and he was glad to be dressed professionally. As Briggs was introduced to others, he was sure to make eye contact, smile, and properly shake hands. Briggs also collected many business cards while networking. During the meal, he was careful to follow dining etiquette. At work the next day, Briggs immediately wrote a thank-you note to the managers who invited him to the event. At the end of the day, Briggs's manager called him to let him know what a great impression he made at the conference. Several colleagues had mentioned to Briggs's manager how impressed they were with his professionalism. Briggs realized that conducting a little research and being professional was well worth the effort.

A common activity in business involves attending social functions. Many invitations request an RSVP, which is French for *répondez s'il vous plaît* (i.e., please respond). As soon as you receive an invitation, send a reply—whether it is an acceptance to attend or a regret that you cannot attend. Not acknowledging the invitation and failing to respond is rude.

When you attend a social function, remember that you are going to network with other professionals. Do not focus on the food; focus on the networking opportunities.

- As with dining situations, refrain from or limit the consumption of alcohol.
- If you choose to eat, serve yourself a small plate of hors d'oeuvres and move away from the food table.
- Hold your hors d'oeuvres in your left hand, leaving your right hand free to shake hands and greet others.
- Do not talk with food in your mouth.
- If there are name badges, wear one placed neatly on your right shoulder.
- If you must handwrite your own name badge, print your first and last name clearly.

Other Etiquette Basics

Research the web for a business etiquette quiz and rate your workplace manners.

At first glance, business etiquette can be a bit overwhelming. However, with practice, business etiquette becomes habit. When in doubt, mimic what the most polished person in the room is doing. Be aware of your surroundings and watch and learn from those whom you admire. The following is a final list of etiquette tips to assist you in becoming one of the most admired and respected individuals in the workplace.

- *Have a pleasant attitude.* In addition to saying "please" and "thank you," do not underestimate the value of a simple smile. A positive attitude will be reflected in your demeanor. When encountering people in the hallways, elevators, and/or meeting rooms, smile, make eye contact, and greet them.
- *Knock before entering an office.* Do not enter an office or private workspace such as a cubicle until you are invited. If the door is open but the individual is with someone else, politely wait your turn. If the

individual you want to see normally has his or her door open, do not disturb the individual when the door is closed. The exception is for an emergency or urgent situation requiring attention, but apologize for the interruption.

- *Put others first.* When you are with colleagues and you are taking turns (in line, to order, etc.), allow your colleagues to go first. Doing so shows respect and courtesy.
- *Apologize when necessary.* Everyone is human and makes mistakes. When you realize that you may have said or done something hurtful to someone, apologize immediately. As you will learn in Chapter 12, apologizing is not a sign of weakness, it is a sign of strength and maturity.
- *Do not use profanity.* The use of profanity is not appropriate in the workplace. Even if others in your presence use profanity, do not assume everyone is comfortable with bad language. Conversations should be professional, respectful, and free of profanity. If you accidentally use inappropriate language, immediately apologize.
- *Avoid dominating a conversation.* There is a key to carrying on a successful conversation: listen. As detailed in Chapter 9, when you are an active listener, you value the information the other individual is providing. Too frequently, individuals dominate a conversation with personal accounts. In general, this is not appropriate because the conversation is focused on you. Next time you are in a conversation, listen to how many times you state the words *me, I,* and *my.* Try to minimize the use of these words in your conversation.

MyStudentSuccessLab   Please visit **MyStudentSuccessLab:** Anderson|Bolt, Professionalism Skills for Workplace Success, 4/e for additional activities, resources, and outcomes assessments.

Workplace Dos and Don'ts

Do wear professional clothes to work	*Don't* wear sweats, flip-flops, or suggestive apparel at work
Do shower and make sure you are always clean	*Don't* overdo cologne or other body sprays
Do make eye contact and offer a gentle but firm handshake	*Don't* grasp only the fingers when shaking hands
Do follow formal dining etiquette at work-related functions	*Don't* reach, grab, or overload your plate at the hors d'oeuvres table
Do say "please" and "thank you" when appropriate	*Don't* assume that the other person knows you are thankful for his or her act of kindness

Concept Review and Application

You are a Successful Student if you:

- Identify appropriate and inappropriate dress for specific workplace situations
- Demonstrate networking behavior, including initiating a professional introduction and handshake
- Apply proper etiquette in a formal dining situation

Summary of Key Concepts

- Projecting an executive presence is important in demonstrating knowledge of basic workplace behavior.
- The majority of first impressions are made through visual appearances.
- Both your maturity and the importance you place on your job are reflected in the way you behave and dress at work.
- Begin to create a work wardrobe today.
- Tattoos, piercings, and body jewelry are offensive to some individuals and are not appropriate in a professional work environment. Consider the long-term consequences of getting a tattoo or piercing.
- Follow business etiquette protocol and consistently utilize it in all areas of your life.
- Make a habit of thanking individuals either verbally or in writing.
- Appropriate etiquette at social functions and while dining is as important as professional behavior at work.

Self-Quiz MATCHING KEY TERMS

Match the key term to the definition using the identifying number.

Key Terms	Answer	Definitions
Casual workdays		1. A standard of social behavior
Courtesy		2. Workdays when companies relax the dress code policy
Dress code		3. Meeting and developing relationships with individuals outside one's immediate work area; the act of creating professional relationships
Etiquette		4. Having the attitude of an executive by demonstrating appropriate workplace behavior
Executive presence		5. An organization's policy regarding appropriate workplace attire
Networking		6. Clothes that are primarily worn to work and work-related functions
Respect		7. Exercising manners, consideration, and respect toward others
Work wardrobe		8. Holding someone in high regard

Think Like a Boss

1. One of your employees comes in on a Monday morning with a pierced tongue and purple hair. What should you do?

2. You have just hired a new employee who clearly has no concept of business etiquette. What specific steps would you take to teach your new employee how to behave professionally?

Activities

Activity 4.1

Research the appropriate attire for your target career and basic wardrobe requirements. Assume you are limited to a $75 budget. Make a list of items you need and could buy to get you through your first week of work. Include the cost.

What You Need to Buy	Cost
	$
Total Cost	**$75**

How will this information affect your personal budget and life plan?

_____

_____

_____

_____

_____

Activity 4.2

Visit a (non-fast-food) restaurant to observe and practice proper dining etiquette. Identify and list inappropriate behaviors others are exhibiting.

Inappropriate Behavior	Why Behavior Is Inappropriate
1.	
2.	
3.	
4.	
5.	

Activity 4.3

Identify three specific areas of etiquette you need to improve and create a plan by listing specific steps for improvement.

Areas of Improvement	Plan
1.	
2.	
3.	

5

Ethics, Politics, and Diversity

fairness • integrity • behavior

After studying these topics, you will benefit by:

- Summarizing how ethics influence personal and professional behavior
- Defending the importance of maintaining confidentiality
- Applying ethical decision making
- Deciding how to respond to unethical behavior
- Classifying the various forms of workplace power and its appropriate use
- Explaining the appropriate use of workplace politics and reciprocity
- Considering the various elements of workplace diversity and its influence on performance
- Stating basic employee rights and legal protections available for workplace diversity issues
- Recognizing the dangers of stereotyping, prejudice, and not respecting cultural differences

HOW DO YOU RATE?

How strong is your moral character?	Yes	No
1. If your new boss openly stole small office supplies, would you report him or her?	☐	☐
2. If your coworkers bad-mouthed another coworker, would you openly defend him or her?	☐	☐
3. If an outside vendor accidentally undercharged your company, would you bring the error to the vendor's attention?	☐	☐
4. If your human resource department overpaid you on your paycheck, would you bring the error to their attention?	☐	☐
5. If company executives asked you to not fully disclose a product defect to customers, would you inform the customer of the defect anyway?	☐	☐

▶ If you answered "no" to two or more of these questions, take a moment to reevaluate who guides and/or influences your personal values. Workplace ethics begin with you. How you conduct business is a reflection of your moral character.

Ethics, Politics, and Diversity at Work

Business is based on competition which sometimes clouds an individual's judgment. Therefore, employees need to recognize how personal values and morals affect ethical behavior at work. Power and politics are used in workplace relationships. Personal ethics govern the outcome of how you deal with power and politics. Although workplace politics are inevitable, it is important that control mechanisms be used appropriately. This chapter focuses on the link between ethics and power, then leads into a discussion of diversity and the influences diversity has in the workplace. Diversity comes in many forms, all of which must be respected and ideally harnessed into a competitive advantage.

Ethics Defined

Throughout our schooling, we are told to behave ethically and do what is right. In education, ethics typically refers to not cheating. Cheating also occurs at work. From the time we clock in to the time we leave the office—and even extending into the weekend—we should behave in an ethical manner. Ethical behavior is a 24-hour process. Personal behavior reflects ethical values. In turn, our ethical behavior reflects and represents our company.

Ethics is a moral standard of right and wrong. Although the definition of ethics is a simple statement, it is important to identify who and what determines what is morally right and wrong. Just as your personality is shaped by outside influences, so is your ethical makeup.

Ethical behavior is a reflection of the influences of friends, family, coworkers, religion, and society. For example, if you associate with people who shoplift, you most likely will not view shoplifting as an unethical act. As a result, you may shoplift without remorse. If your family routinely lies about a child's age to pay a lower admission fee to a movie theater or amusement park, the child is being taught to be dishonest. Common religions teach that lying, cheating, and stealing are wrong. Consider the influences our society and culture have on our ethical behavior. Corporate America has been bombarded with ethics-related scandals. Additionally, many of the messages we receive on a daily basis influence our perception of what is and is not ethical behavior.

Although the preceding factors all have an enormous influence on the makeup of one's ethics, note that ethical behavior starts with the individual. As we explore the concept of ethics at work, remember that ethics begin with you.

Ethical behavior is the foundation of creating trust. Both ethics and trust are crucial elements of leadership. Although we will discuss leadership in a future chapter, it is important to make the link between ethical behavior and leadership. We often hear people state that more business leaders should have moral character; what does this really mean? As stated earlier, ethics are a moral standard of right and wrong as defined by society, whereas **morals** are a personal standard of right and wrong. An individual's **values** are the important beliefs that guide his or her behavior. Both morals and values guide and influence behavior at work and determine character. **Character** is defined as the unique qualities of an individual and is usually a reflection of personal morals and values. Therefore, if someone says an individual is of high character, he or she is most likely stating that the individual is honest and fair. **Integrity** is when

Talk It Out

What unethical behaviors do students display during class?

Talk It Out

Discuss recent corporate ethics-related scandals.

someone consistently behaves in an ethical manner. Strive to display integrity when unethical situations arise.

In Chapter 1, we mentioned the link between ethics and values and how values affect character. In a competitive work setting, situations are often stressful and filled with hidden agendas and underlying politics. Maintain integrity and do not be tempted to give in to negative behavior. How you respond to a situation reveals your character. Just as with first impressions, once someone has formed an opinion of you and your character, it is difficult to change that opinion. Therefore, ensure your character consistently projects honesty and fairness and reflects your personal morals and values.

Think About It

How do ethics, values, character, and integrity factor into your life plan?

Values, Conflicts, and Confidentiality

Each company has a corporate culture. A corporate culture is the way a company's employees behave. It is based on the behavior of its leaders. For example, if all the executives within the company are very laid-back and informal, employees throughout the company will most likely be laid-back and informal. If a department supervisor is stressed and unprepared, the department members will most likely be stressed and unprepared. A company's corporate culture is also a reflection of an organization's attitude toward ethical behavior. Many companies promote ethical behavior through an **ethics statement**, which is a formal corporate policy that addresses the issues of ethical behavior and punishment should someone behave inappropriately. As corporate America recovers from its scandals, more companies are placing greater importance on ethics statements. Included in most corporate ethics policies is a statement addressing conflicts of interest. A **conflict of interest** occurs when you are in a position to influence a decision from which you could benefit directly or indirectly.

Topic Situation

Nancy's company needs a flower vendor for an upcoming event. Nancy's uncle owns a local flower shop, and getting this contract would be a big financial boost to his store. Nancy wonders if it would be unethical to tell her uncle about the opportunity.

TOPIC RESPONSE
Name several actions Nancy can take to determine if this is a conflict of interest.

If there is ever a possibility that someone can accuse you of a conflict of interest, excuse yourself from the decision-making process. If you are uncertain whether there is a conflict, check with your boss or the respective committee. Explain the situation and ask for your boss's or committee's opinion. To avoid a conflict of interest, many companies have strict policies on gift giving and receiving. Many companies do not allow the acceptance of gifts or have a maximum dollar value of a gift allowed.

Certain work-related information and issues are **confidential**, meaning they are private. These matters may include client records, employee information, business reports, documentation, and files. **Implied confidentiality** is when an employee has an obligation to not share information with individuals for whom the business is of no concern. Regardless of whether you have been explicitly told to keep work-related information confidential, practice implied confidentiality.

An example of implied confidentiality is to not share customers' personal information with others unless you are told to do so.

Sometimes you may be tempted or even asked to share confidential information. Do not fall into this trap. If you are uncertain about sharing confidential information with someone, check with your boss. Doing so will demonstrate to your boss that you want to not only maintain the privacy of your department, but also behave in a professional manner.

Exercise 5.1

Your company has a strict policy on not accepting gifts valued over $15. A key vendor for your company sends you flowers on your birthday. The arrangement is quite large, so you know it clearly exceeds the $15 limit. What do you do?

_____

_____

_____

Making Ethical Choices

At work, you will be faced with ethical decisions including the appropriate use of workplace power and politics. Consistently practicing ethical behavior at work is not always easy. As you attempt to make ethical choices, use the three **levels of ethical decisions**, which are three questions to help you make an ethical decision. Each level contains a question that, if answered "no," deems the decision unethical.

1. *First level:* Is the action legal? When confronted with an ethical issue, first ask if the action is legal. If the answer is no, the action is illegal; therefore, the action is unethical.

2. *Second level:* Is the action fair to all involved? If the answer is no, the action is not fair to all involved; therefore, the action is unethical. Your behavior should be fair to all parties involved. If, when making a decision, someone is clearly going to be harmed or is unable to defend him- or herself, the decision is probably not ethical. Note that the concept of fairness does not mean that everyone is happy with the outcome. It only means that the decision has been made in an impartial and unbiased manner. Although a behavior may be legal, it still may be considered unethical. Just because a behavior is legal does not mean it is right. Take the case of an individual who has a romantic relationship with someone who is married to someone else. While in some states it is not illegal to have an extramarital affair, this behavior is unethical.

3. *Third level:* Does the behavior make you feel good? It is understandable that not everyone agrees on what is right and fair. If the answer to this question is no and you are not feeling good about the behavior, it is most likely unethical. This is when the ethical decision gets personal because your conscience becomes a consideration. In the classic Disney movie *Pinocchio,* the character Jiminy Cricket was Pinocchio's conscience. He made Pinocchio feel bad when Pinocchio behaved inappropriately. Just like Pinocchio, each individual has a conscience.

When people knowingly behave inappropriately, most will eventually feel bad about their poor behavior. Some people take a bit longer than others to feel bad, but most everyone at some point feels bad when they have wronged another. Sometimes a behavior may be legal and it may be fair to others, but it still may make an individual feel guilty or bad.

Exercise 5.2

Autumn is responsible for a petty cash box at work. Autumn is planning on going to lunch with friends but does not have time to stop by the ATM until later in the afternoon. Autumn struggles with the thought of temporarily borrowing $10 from the petty cash box and returning the money later in the day (after a visit to the ATM). No one would ever know. Technically, it is not stealing, it is just borrowing. As Autumn debates temporarily borrowing the money, Autumn begins to feel guilty. Autumn decides the behavior is unethical, does not take the petty cash, and skips going out to lunch with friends.

Based on Autumn's dilemma, answer the three levels of ethical behavior.

1. Is Autumn borrowing money from the petty cash box legal? _____

2. Is Autumn borrowing the money fair? _____

3. Would you feel bad about borrowing the money? _____

Exercise 5.3

It is 9:00 P.M., it is raining, and you are hungry. You are on your way home from a long workday. You only have $5 in your wallet, so you decide to go to a fast-food drive-through to get dinner. You carefully order so as not to exceed your $5 limit. You hand the drive-through employee your $5, and he gives you change and your meal. You discover that the fast-food employee gave you change for $20. What do you do? Apply the three levels of ethical decision making to this scenario.

1. Is it legal to keep the money? _____

2. Is it fair to keep the money? _____

3. Do you feel bad about keeping the money? _____

Exercise 5.4

Typically, in the fast-food business, employees whose cash drawers are either short or over cash more than once are at risk of being fired. If you initially were going to keep the money from Exercise 5.3, but now you know the employee who gave you too much cash could get fired because you decided to keep the money, would you still keep it?

_____

_____

_____

Workplace Power

Power is one's ability to influence another's behavior. Whether you recognize it or not, everyone at work has and uses some form of power. Successful employees understand this ability to influence others' behavior and use it appropriately. Knowing the different types of power and how to properly use each type at work will increase your ability to influence others' behavior. The seven bases of power include legitimate, coercive, reward, connection, charismatic, information, and expert power.

Legitimate power is the power that is given to you by the company. It includes your title and any other formal authority that comes with your position at work. For example, a manager has legitimate power to assign schedules for employees in his or her department. **Coercive power** is also power that is derived from your formal position. However, coercive power is negative and uses threats and punishment. An example of coercive power is if your manager threatens to cut your hours. A contrast to coercive power is **reward power**. Reward power is the ability to influence someone with something of value. For example, a manager may offer you a bonus for meeting a goal. Those with legitimate power can reward others with promotions, pay increases, and other incentives. However, you do not have to have legitimate power to reward others in the workplace. This will be further explained in the next section.

Connection power is based on using someone else's legitimate power. Consider the department assistant who arranges meetings based on his or her boss's power. The department assistant has a connection to an individual with authority. The three remaining types of power are often referred to as *personal power* because they come from within. **Charismatic power** is a form of personal power that makes people attracted to you. Most of us know someone who walks into a room and immediately attracts other people's attention. This is because the individual with charismatic power or charisma shows a sincere interest in others. **Information power** is based on an individual's ability to obtain and share information. Doing so makes you more valuable to those with whom you interact. For example, a coworker who is part of a committee routinely shares information from the committee meetings. **Expert power** is power that is earned by one's knowledge, experience, or expertise. Consider the company's computer repair technician. On the company's organization chart, he or she may not be very high in the formal chain of command. However, this individual wields a lot of power because of his or her computer expertise.

Exercise 5.5

Identify three ways employees without legitimate power can reward others.

_____

_____

_____

Increasing Your Power Bases

As mentioned earlier, everyone possesses some form of workplace power. The trick is to recognize, utilize, and increase your power. The easiest way to increase legitimate power is to make people aware of your title and responsibilities. When doing so, be cautious not to brag or act conceited. Coercive power should be used only when an individual is breaking policy or behaving inappropriately. Even then, only use coercive power in a confidential and respectful manner.

Reward power can and should be used daily. Whenever possible, dispense a sincere word or note of appreciation to a coworker who has assisted you or has performed exceptionally well. Doing so will develop and enhance relationships not only within your department, but also outside your department. Be sincere. Increase your connection power by strengthening your network. As discussed in Chapter 4, networking means meeting and developing relationships with individuals outside your immediate work area. Network with individuals within and outside of your organization. Try to make at least one new professional contact a week. When interacting with others, make eye contact, initiate a conversation, and focus the conversation on the other individual instead of on you. Increasing charismatic power in this way establishes trust, improves communication, and makes others want to work with you. Information power is also developed by attending meetings, joining committees, and networking. Whenever possible and without overcommitting, join committees, attend meetings, and actively share information with appropriate individuals. Doing so exposes you to other people and issues throughout the company. You, in turn, not only learn more about what's going on within the organization, but you also increase your connection power. Increase your expert power by practicing continual learning. Read books and business-related articles, scan reputable and applicable Internet sites, attend workshops and conferences, and learn new technology when possible. Whenever you learn something new that can assist others at work, share this information. Coworkers will see you as the expert in this area.

Workplace Politics and Reciprocity

When you begin to obtain and utilize workplace power, you are practicing politics. **Politics** is obtaining and using power. People generally get a bad taste in their mouth when someone accuses them of being political, but this is not necessarily a bad thing. As mentioned earlier, recognizing, increasing, and appropriately utilizing the various power bases at work is encouraged. It is when one expects reciprocity that politics at work get dangerous. **Reciprocity** is when debts and obligations are created for doing something. Suppose you are on a time crunch and must complete a report in two hours. You ask a coworker to help you. The coworker stops what he or she is doing and assists you with an hour to spare. You have just created a reciprocal relationship with the coworker. When this coworker is in a crunch, he or she will not only ask you, but expect you to help. The workplace is comprised of reciprocal relationships. Unfortunately, sometimes the phrase "you owe me" is used by a coworker who may try to encroach on your ability to behave ethically.

TOPIC RESPONSE
What may happen if Ryan does lie for his coworker?

Ryan has a coworker who helps him with special projects when time is short. When the coworker tells Ryan she needs help with something, Ryan immediately responds, "Sure, no problem." Unfortunately, there is a problem: The coworker wants Ryan to attend a meeting for her and tell those at the meeting that she is home sick when Ryan knows she plans to take a trip with friends. Ryan tells the coworker that it would be unethical to cover for her. "But you owe me!" says the coworker. Ryan is unsure what to do. After some thought, Ryan tells the coworker that he wants to repay the favor and appreciates all the help the coworker provides, but he will not lie for her. Ryan should expect some tension with this coworker, but in the long run, she will respect Ryan.

When Others are Not Ethical

The previous section discussed how to behave ethically at work. But what should you do when others are not behaving ethically? Let us review the three levels of ethical decision making. Everyone must abide by the law. If someone at work is breaking the law, you have an obligation to inform your employer immediately. This can be done confidentially to either your supervisor or the human resource department. Before you accuse anyone of wrongdoing, have documentation that presents the facts. Keep track of important dates, events, and copies of your evidence and documentation. Your credibility is at stake. If the offense is extreme and obviously illegal, alert outside officials. If the offense is not extreme and you are uncertain if the offense is illegal, contact management and seek their advice on next steps. If management tolerates the behavior, you must accept management's decision; however, if you are still bothered by the behavior and feel it is inappropriate, then determine if you want to continue working for the organization.

TOPIC RESPONSE
Do you believe it was fair for the coworker to keep the smartphone? Why or why not?

Tony finds out that a certain coworker received a smartphone from a vendor for personal use. No other employee received a smartphone. The coworker said the device was an incentive for the company's good standing with the vendor and, because he was the employee who made the purchases, it was his right to keep the device. Tony thinks this is unfair and unethical. Tony politely checks with the human resource department, and they tell Tony that the coworker can keep the smartphone. Although Tony construed the situation as unethical, the company found no conflict with its policies. Tony decides the offense is not extreme, and because it was accepted by management, Tony accepts management's decision.

Common Ethical Issues

Another common ethical issue at work occurs in the area of company theft. Company theft is not always large items such as laptops or other equipment. More often, it is smaller items such as office supplies. You may not realize that taking a pen or pencil home is also stealing from your company. Work supplies should be used only for work purposes.

Ethical behavior is reflected in your dependability and how you conduct yourself on company time. The company is paying you to perform your job. If you commonly use company time for personal business, you are stealing time from the company. Although at times it may be necessary to conduct personal business during work hours, it is inappropriate to consistently spend your time on non-work-related activities. The following activities should not be conducted during work hours:

- Using the Internet for personal business such as social networking or shopping.
- Taking or making personal calls or texts.
- Exceeding allotted break and lunch periods.
- Playing computer games.
- Using company supplies and equipment for non-business purposes.

If you must conduct personal business while at work, do so only during your break or lunch hour. Whenever possible, conduct personal business before or after work hours and in a private manner.

Talk It Out

When is it appropriate to make personal calls at work?

Diversity Basics

In 2013, the Internal Revenue Service (IRS) was accused of unfairly targeting certain individuals based on their personal beliefs. This event provides an unfortunate example of how ethics, politics, and diversity are interrelated. Diversity comes in many forms. Although most people think of diversity as a race issue, the topic goes far beyond race. People are different in many aspects, ranging from ethnicity, to political and religious beliefs, to the way we wear our hair. **Workplace diversity** refers to differences among coworkers. These differences include age (generational differences), gender, economic status, physical makeup, intelligence, religion, and sexual orientation, among other things.

As we discuss these issues, note three primary messages regarding workplace diversity:

- No matter how we differ, everyone should be treated fairly, with respect, and with professionalism.
- Diversity should be viewed as an asset that utilizes our differences as a means to create, innovate, and compete as an organization.
- Individual differences related to diversity should only be an issue when the diversity negatively affects performance.

Topic Situation

One of Dianne's new friends at work has a different religion than hers. Although Dianne's friend has never openly mentioned his religious beliefs, the new friend subtly slips into his office at routine times of the day and eats a special diet. Dianne really enjoys the workplace friendship with this coworker and, although Dianne is confident in her own beliefs, Dianne wants to learn more about the friend's beliefs. Dianne sometimes does not know how to behave around this friend when the topic of meals or religion comes up. Dianne thinks about talking to the coworker about the differences but does not want to offend him.

TOPIC RESPONSE

Is it appropriate for Dianne to ask her coworker about a religious difference? If no, why? If yes, how should she phrase her question?

Legal Protection from Discrimination

The Equal Employment Opportunity Commission (EEOC) enforces laws that protect individuals from workplace discrimination in recruiting, hiring, wages, promotions, and unlawful termination. These laws are based on Title VII of the Civil Rights Act, which prohibits discrimination based on sex, religion, race or color, or national origin. Since that time, additional laws have been made to further protect individuals from discrimination in the areas of age (40 and older), physical and mental disabilities, gender, sexual orientation, hate crimes, pregnancy, and military service. Individuals that fall into one of the previously mentioned categories are members of a **protected class**, a group of individuals who are protected from discrimination based on civil rights legislation. **Workplace discrimination** is acting against someone who is a member of a protected class.

If you ever feel you are a victim of workplace discrimination, first contact your human resource department. Should you feel you are still experiencing discrimination, contact your state's Department of Fair Employment and Housing or Department of Labor (depending on your state) or the Equal Employment Opportunity Commission. Every employee has a right to a workplace that is free from discrimination.

Race is defined as people having certain physical traits. Racial differences include various ethnicities, such as Hispanic, Asian, African American, Native American, and Anglo-Saxon. **Culture** is the different behavior patterns of people. Examples of various cultures may include where you live geographically, your age, your economic status, and your religious beliefs. As the workplace becomes more diverse, it is hard to imagine a workplace that does not include various races and cultures.

As we move into a global economy, recognize and respect global differences. Americans sometimes are accused of being **ethnocentric**. Ethnocentrism is when an individual believes his or her culture is superior to other cultures. It seems a bit foolish when Americans behave in an ethnocentric manner because America is really a compilation of many cultures. No one culture is better than another. In the workplace, respect and be aware of how various cultures affect the work environment.

Understanding how race and culture influence our workplace helps us begin to recognize how these differences influence our values and behavior. In Chapter 1, we discussed that not everyone thinks and behaves like you. Moreover, people look different and have different value systems. Although we may not like someone's looks or agree with others' values or religious beliefs, we must respect everyone's differences and treat them professionally.

Digging deeper into the issue of culture, we need to appreciate various generational differences and their impact on the workplace. Individuals entering the workforce (18 to 22-year-olds) have different needs than those preparing to retire (55 and older). Moreover, these needs reflect different priorities, values, and attitudes regarding leadership, workplace relationships, and the use of technology.

Think About It

Look around the room and identify differences between you and your classmates.

Stereotypes and Prejudice

In Chapter 1, we discussed the differences in people's attitudes and how these attitudes form our personalities. Everyone is a product of past experience. Individuals use these past experiences to form perceptions about people and situations. A **perception** is one's understanding or interpretation of reality. If we had a positive previous experience, we will most likely have a positive perception of a person or circumstance. For example, if your boss calls you in to his or her office, you will either have a positive or negative perception of the impending conversation. If your boss is a good communicator and you frequently visit his or her office, you will have a positive perception of being called in to the office. On the other hand, if your boss calls you in to his or her office only for negative purposes, your perception of reality is that the boss's office solely represents reprimands and punishment.

To make situations easier to understand or perceive, we often stereotype. **Stereotyping** is making a generalized image of a particular group or situation. Often, our perceptions mold how we view groups or situations. These images can be positive or negative, but we generally apply them to similar situations and groups. At work, this can include types of meetings (situations) or members of specific departments (groups). Using the preceding example of the boss and his or her office, if you have a positive perception of your boss, you could most likely stereotype that all bosses are good communicators.

It is wise to not only know the definition of stereotyping, but also to avoid applying stereotypes in a negative manner. For example, a common stereotype is that athletes are not intelligent. This is simply not true. Prior to responding to a situation, conduct an attitude check to ensure that you are not basing your reaction on a perception or stereotype rather than responding to the current facts and situation.

Using the previous example, if we assume that all athletes are unintelligent (stereotype), we have demonstrated prejudice. **Prejudice** is a favorable or unfavorable judgment or opinion based on one's perception (or understanding) of a group, individual, or situation. Typically, at work, prejudice is a negative attitude or opinion that results in discrimination. Therefore, if we do not hire athletes because we believe they are not intelligent, we are guilty of discrimination. Many people harbor some form of prejudice. Recognize in what areas you may be harboring prejudice and begin understanding why. Once you recognize what areas need improvement, begin taking action to decrease your prejudice. One way is to learn about the individual, group, or situation that is causing the prejudice.

Exercise 5.6

What areas of prejudice do you see on campus? What areas of prejudice do you see in your community?

_____

_____

_____

Labeling is when we describe an individual or group of individuals based on past actions. We attach positive or negative labels to groups or individuals, and then watch for supporting behaviors to see if these behaviors live up to or dispel our labels. For example, if we label a coworker as being the smartest person we know, that person may live up to this expectation by behaving as the smartest person (regardless of whether he or she really is intelligent). However, he or she may dispel the label by purposely behaving ignorantly.

Assumptions sometimes are made at work based on people's language differences and accents. These assumptions may include economic status, intelligence, and social customs. In our melting pot society, it is common for individuals to speak a different language when at home. At work, speaking a second language can be a means of attracting and meeting customer needs. Therefore, being bilingual can be a workplace asset.

Do not make fun of people with different cultures or lifestyles or individuals with physical and mental disabilities. Even jokes that we believe are innocent may cause deep wounds. Moreover, they may not only be offensive, but may violate someone's civil rights. Inappropriate comments can be construed as both workplace discrimination and harassment.

Topic Situation

TOPIC RESPONSE

What are Gurbinder's options in handling this situation? What is the best option? Why?

Gurbinder is invited to lunch with some new coworkers. During the meal, one of the coworkers tells a joke about a blind person of a certain ethnicity. Gurbinder politely chuckles at the punch line but is actually offended. Gurbinder wonders how to best handle the situation. Should he tell the joke teller that the joke was offensive? Should he tell the department supervisor? Gurbinder clearly believes that this type of behavior is inappropriate.

Companies are attempting to better address workplace diversity through several actions. First, they are developing **diversity statements**. These are statements that remind employees that diversity in the workplace is an asset and to not engage in prejudice and stereotyping. Second, companies are providing diversity training to teach employees how to deal with and ultimately eliminate workplace discrimination and harassment. This training commonly applies to all employees, customers, and vendors. Third, companies are eliminating the **glass ceiling** and **glass walls**. Glass ceilings and glass walls are invisible barriers that frequently make executive positions (glass ceiling) and certain work areas or work-related places (glass walls) off limits to those in a protected class. Proactive companies offer formal mentoring programs to assist in identifying and training those in a protected class for promotion opportunities. People should not receive special treatment because they are members of a protected class, but they should be given an equal opportunity. The employer is responsible for hiring the most qualified candidate.

Cultural Differences

Our society is a mix of individuals from all over the world. For this reason, it is important to address cultural differences and their impact on the workplace. Cultural differences may include religious influences and are related to the treatment of individuals based on age, gender, and family.

There are many different religions in the world. Although most major U.S. holidays are based around Christian holidays, not everyone who works in the United States is a Christian. Individuals who do not share your religious values are afforded the same rights as you. As mentioned earlier in this chapter, the Civil Rights Act protects individuals from discrimination based on religion. Everyone is entitled to observe his or her respective religious holidays and traditions. Once again, we must be respectful of everyone's individual religious beliefs and not condemn someone for his or her religious difference. Although an individual's religious beliefs may permeate every element of his or her life, as with other issues of diversity, if religion negatively affects performance, the issue must be addressed.

Some countries have cultures that focus on the individual, whereas other countries prioritize what is best for society over personal needs. We should recognize and respect cultural differences and not offend others. For example, some hand gestures commonly used in the United States may be offensive to someone who has come from another country. If you feel you may have offended someone based on a cultural difference, find out what behavior offended the other person, apologize if necessary, and try to not repeat the offensive behavior.

Cultural differences can have both a positive and negative impact on business. Learning about other cultures can provide insights into new markets and stimulate creativity. With so much diversity among employees and customers, knowing other cultures will result in improved relationships. Outcomes can be negative when companies do not properly train employees and address cultural differences; this is when prejudice and discrimination may emerge.

MyStudentSuccessLab | Please visit **MyStudentSuccessLab:** Anderson|Bolt, Professionalism Skills for Workplace Success, 4/e for additional activities, resources, and outcomes assessments.

Workplace Dos and Don'ts

Do consistently behave in an ethical manner	*Don't* behave one way at work and another around your friends
Do keep information confidential	*Don't* break the company's trust
Do recognize and increase your workplace power bases	*Don't* use your workplace power in a harmful or unethical manner
Do know your rights regarding workplace diversity	*Don't* accept defeat in discriminatory situations
Do learn to respect differences in others	*Don't* use your minority status to take advantage of situations
Do be proud of your culture and heritage	*Don't* show prejudice toward others
Do take responsibility for increasing awareness about workplace diversity issues	*Don't* label people

Concept Review and Application

You are a Successful Student if you:

- Explain how consistently making ethical choices creates a foundation for professional success toward the goals you established in your life plan

- Name and define the advantages and appropriate use for each power base

- State basic rights and legal protection employees have from workplace discrimination

Summary of Key Concepts

- Personal ethical behavior is a reflection of the influences of friends, family, religion, and society.

- Do not share confidential information with individuals for whom the business is of no concern.

- Power and power bases are effective tools to use in the workplace.

- Be cautious to not use power and reciprocity in an unethical manner.

- A conflict of interest occurs when you are in a position to influence a decision from which you could benefit directly or indirectly.

- No matter what our differences, treat everyone with respect and professionalism.

- Title VII of the Civil Rights Act prohibits discrimination based on sex, religion, race or color, or national origin.

- Diversity should be used as an asset to create, innovate, and compete.

- Workplace diversity should be an issue only when the diversity negatively affects performance.

Self-Quiz MATCHING KEY TERMS

Match the key term to the definition using the identifying number.

Key Terms	Answer	Definitions
Character		1. Personal power that makes people attracted to you
Charismatic power		2. Power that is earned by one's knowledge, experience, or expertise
Coercive power		3. When an individual believes his or her culture is superior to other cultures
Confidential		4. Society's moral standard of right and wrong
Conflict of interest		5. Power based on an individual's ability to obtain and share information
Connection power		6. Important beliefs that guide an individual's behavior

Key Terms	Answer	Definitions
Culture		7. Three questions to help an individual make an ethical decision
Diversity statements		8. The unique qualities of an individual
Ethics		9. Using someone else's legitimate power
Ethics statement		10. Obtaining and using power
Ethnocentric		11. When debts and obligations are created for doing something
Expert power		12. Personal standard of right and wrong
Glass ceiling		13. Describing an individual or group of individuals based on past actions
Glass wall		14. Power that is given to you by the company
Implied confidentiality		15. One's understanding or interpretation of reality
Information power		16. Statements that remind employees that diversity in the workplace is an asset and not to engage in prejudice and stereotyping
Integrity		17. Negative power that uses threats and punishment
Labeling		18. Different behavior patterns of people
Legitimate power		19. The ability to influence someone with something of value
Levels of ethical decisions		20. A favorable or unfavorable judgment or opinion toward an individual or group based on one's perception
Morals		21. Matters that are private
Perception		22. Groups of individuals who are protected from discrimination based on civil rights legislation
Politics		23. Differences among coworkers
Power		24. An employee's obligation to not share information with individuals for whom the business is of no concern
Prejudice		25. Making a generalized image of a particular group or situation
Protected class		26. When someone consistently behaves in an ethical manner
Race		27. Acting against someone who is a member of a protected class
Reciprocity		28. A position to influence a decision from which you can benefit directly or indirectly
Reward power		29. One's ability to influence another's behavior
Stereotyping		30. Invisible barriers that frequently make executive positions off-limits to those in a protected class
Values		31. Invisible barriers that frequently make certain work areas or work-related places off-limits to those in a protected class
Workplace discrimination		32. A formal corporate policy that addresses the issues of ethical behavior
Workplace diversity		33. People having certain physical traits

Think Like a Boss

1. What is the best method of dealing with an ethical decision regarding the performance of an employee?

2. What would you do if you noticed an employee treating another employee in a discriminatory manner?

3. What can you do to minimize workplace discrimination and harassment?

Activities

Activity 5.1

The term *integrity* is derived from the root word *integer*. Research both terms and write a brief summary of how these terms relate to the appropriate use of workplace power and politics.

_____

_____

_____

Activity 5.2

Identify a time you overheard confidential information that should not have been shared—for example, sitting in a physician's office or overhearing a private conversation while shopping.

_____

_____

How should this situation have been better handled?

_____

_____

Activity 5.3

Identify at least three potential areas where workers might be tempted to be dishonest or breach confidentiality.

_____

_____

_____

Activity 5.4

Is it appropriate to discuss the following company information with individuals outside of the company? Why or why not?

Information	Appropriate (Yes or No)	Why or Why Not?
1. Key clients/customers		
2. Financial information		
3. Boss's work style		
4. Company mission statement		
5. Names of members of the company board of directors		

Activity 5.5

Research a country to identify common workplace practices (e.g., gender, religion, attire, gift giving, and meetings). Report your findings.

_____

_____

_____

Share what this activity taught you about diversity.

_____

_____

Activity 5.6

Identify a recent experience where you observed an act of prejudice.

_____

_____

How could you have handled the situation differently?

_____

_____

6

Accountability and Workplace Relationships

responsible • respectful • involved

After studying these topics, you will benefit by:

- Defining and linking concepts of empowerment, personal responsibility, and accountability
- Explaining how workplace relationships affect workplace success
- Identifying appropriate and inappropriate relationships with your boss, colleagues, executives, vendors, and customers
- Justifying how best to respond to a negative workplace relationship
- Stating basic expectations regarding work-related social functions, situations, and office issues

HOW DO YOU RATE?

Who cares?	True	False
1. I should not have to clean the break room if I did not make it dirty.	☐	☐
2. I should not have to unjam the copy machine if I was not the one who jammed it.	☐	☐
3. I have the right to date anyone I please, including coworkers, vendors, and customers.	☐	☐
4. I have the right to privately and discreetly share my dislike of my boss or coworker with other employees if they promise to keep the matter confidential.	☐	☐
5. I do not have to attend company functions if I don't want to attend.	☐	☐

▶ If you answered "true" to two or more of these questions, you are correct, you have the right to make workplace choices on these sensitive matters. However, your actions should reflect accountability and responsibility. Consistently respond to situations such as these in a manner that reflects positively on both you and your employer.

Accountability and Empowerment

Employees need to be accountable—but what does that really mean? Accountability involves taking personal responsibility for ensuring your work reflects positively on both you and your company. This may include a project, a customer experience, and workplace and vendor relationships. In politics, business, and education, individuals need to be held accountable for their actions. Unfortunately, too many people do not demonstrate accountability. This chapter discusses the concepts of accountability and workplace relationships. The concepts of empowerment, responsibility, and accountability are all about personal choices. These personal choices not only affect how successfully you will perform at work, but have a tremendous impact on workplace relationships.

In Chapter 5 we discussed power bases and how workplace power affects politics and ethical behavior. All employees have power; unfortunately, many of them do not use their power appropriately or at all. As companies place an increased focus on quality and performance, good decision making by employees becomes more important.

Empowerment is pushing power and decision making to the individuals who are closest to the customer in an effort to increase quality, customer satisfaction, and, ultimately, profits. The foundation of this basic management concept means that if employees feel they are making a direct contribution to a company's activities, they will perform better.

Consider the case of a manager for a retail customer service counter telling his employee to make the customer happy. The manager feels he has empowered his employee. However, the next day, the manager walks by the employee's counter and notices that the employee has given all customers refunds for all returns, even when some returns did not warrant a refund.

The boss immediately disciplines the employee for granting the refunds. The employee made the customer happy, just as the manager had asked. The manager believed he empowered his employee, but he did not.

Telling someone to do something is different than teaching someone the correct behavior. The employee interpreted the phrase "make the customer happy" differently from the manager's intention. The proper way for the manager to have empowered the employee would have been to discuss the company's return policies, role-play various customer scenarios, and then monitor the employee's performance. During the training process, the wrong behavior should have been immediately corrected and good performance should have immediately received positive reinforcement.

When you, as an employee, demonstrate a willingness to learn, you have taken responsibility. **Responsibility** is accepting the power that is being given to you. If you are not being responsible, you are not fully utilizing power that has been entrusted to you. The concept of empowerment and responsibility is useless without accountability. **Accountability** means that you will report back to whoever empowered you to carry out a specific responsibility. Employees at all levels of an organization should consistently perform their best because they are accountable to each other, their bosses, their customers, and the company's investors.

To gain respect and credibility at work, begin asking for and assuming new tasks. If you are interested in learning new skills, speak up and ask your boss to teach or provide you opportunities that will increase your value to the company. Assume responsibility for these new tasks and report back (become accountable) on your performance. Worthwhile activities support the company's overall mission. Each project for which you assume responsibility must have a measurable goal. If your project lacks a measurable goal or outcome, it will be difficult to be accountable for your performance.

As you increase workplace responsibilities, do not be afraid to seek assistance. Learning comes from others and from past experiences. When you make mistakes, apologize if necessary and do not blame others. Determine what went wrong and why. Learn from mistakes and view them as opportunities to perform better in the future.

Topic Situation

Christine is rapidly becoming more confident on the job. Wanting to be of more value to the company, she began studying the concept of personal accountability and requesting extra projects at work. Christine gladly kept the boss informed on the status of each project and reported when each project was successfully completed. The boss noticed how Christine took responsibility not only for personal growth, but also for the success of the department. As a result, the boss informed Christine that he was impressed with her willingness to improve and is considering Christine for a promotion.

TOPIC RESPONSE
What additional activities could Christine perform to demonstrate she is responsible and accountable?

Personal Accountability

In today's uncertain economy, it is critical that employees show up to work with a positive attitude and give 100 percent effort when at work. Companies are continuously exploring methods to increase efficiency and decrease waste. Accountable employees take personal responsibility for their performance, actions, and workplace choices. Be on time and do what is expected of you. Do not miss work for personal pleasure, and do not call in sick unless you truly are ill. Reserve personal leave days for emergencies, funerals, or other such situations. During work hours, work. Do not surf the Internet or waste company time on personal activities. When an employee is constantly late, absent, or not completing his or her duties, other employees must assume the responsibilities of this employee. Not being accountable to your coworkers leads to poor workplace relationships and affects productivity.

Workplace Relationships

People who are mature and confident behave consistently around others, whereas those who are not as secure with themselves frequently behave differently around the boss or selected colleagues. It is wrong and immature to behave inconsistently at work. Behave professionally and respectfully, no matter who is present or watching. This section explains the dynamics of workplace relationships and their impact on performance.

Because many people spend more time at work than at home, workplace relationships have a profound impact on productivity. Treat everyone respectfully and professionally. It is easy to be respectful and professional to those we like;

it is much more difficult doing so with those with whom we do not get along. Chapter 12, which covers the topic of conflict at work, addresses the sensitive issue of working productively with those we do not like. Unfortunately, having strong friendships at work can sometimes be equally as damaging as having workplace enemies if we fail to keep professional relationships separate from our personal lives. Socializing with our coworkers is both expected and acceptable to a degree, but do not make workplace relationships your only circle of friends. Doing so is dangerous because it becomes difficult to separate personal issues from work issues. It also has the potential to create distrust among employees who are not included in your circle of workplace friends. Finally, it creates the potential for you to subconsciously show favoritism toward your friends. Even if you are not showing favoritism, those who are not within the circle of friends may perceive favoritism and may become distrustful of you.

As you become more comfortable with your job and company, you will be in various situations that provide opportunities to strengthen workplace relationships with coworkers, executives, investors, vendors, and customers. The following section addresses selected situations and how best to behave in these circumstances.

Executives/Senior Officials

It is often difficult for employees to know how to behave in a roomful of executives or board members, such as in meetings, corporate events, and social functions. Although it is tempting to pull a senior official aside and tell him or her stories about your boss or how perfect you would be for an advanced position, be aware that some may view this behavior as both inappropriate and unacceptable. Do not draw attention to yourself by appearing overly assertive. When given the opportunity, project a positive, professional image. Highlight the successes of your department instead of personal accomplishments while being humble and gracious.

If you are in a meeting that you do not normally attend, pay attention, sit quietly, and interject your comments only if necessary. Do not dominate the discussion. If it is convenient, network by introducing yourself to senior officials before or after the meeting, but do not interrupt when others are engaged in conversation. Be confident, make eye contact, extend a hand, and state, "Hello, Mrs. Habib, my name is Tim Brandon. I work in the accounting department. It's nice to meet you." Keep your comments brief and positive. Your objective is to create a favorable and memorable impression with the executive. Do not speak poorly of anyone or a situation. It is also inappropriate to discuss personal work-related issues, such as wanting to change positions, unless you are in a meeting specifically to discuss that issue. Allow the executive to guide the conversation. During the conversation read the executive's body language. If the executive's body language includes a nodding head and his or her body is facing you, continue visiting. If the executive is glancing away or his or her body is turning away from you, that body language is communicating the executive's desire to be elsewhere; therefore, politely end the conversation, excuse yourself, and leave. Use networking encounters with executives as opportunities to create favorable impressions for you and your department.

Talk It Out

Why should you not speak poorly of others when networking?

Boss Styles

Many workers have strong emotions about their bosses (sometimes favorable, sometimes unfavorable). Prior to discussing how to professionally work with various types of bosses, we must remember that bosses are human. Like us, they are continually learning and developing their skills. Although they are not perfect, assume they are doing their best.

If you have a good boss, be thankful. A good boss is one who is respectful and fair and will groom you for a promotion. It frequently becomes tempting to develop a personal friendship with a good boss. Keep the relationship professional. It is okay to share important activities occurring in your personal life (e.g., spouse and child accomplishments, vacation plans), but do not divulge too much personal information. Turn the relationship into a mentoring opportunity by learning and developing the management and leadership qualities that you admire in your boss, and begin integrating these qualities in your own workplace behavior.

There may be other situations where we feel our boss is incompetent and appears to not know how to do his or her job. Ask yourself whether the negative perception is based on fact or on your own insecurities. As with any work situation, no matter how bad we perceive our boss to be, remain professional and respectful. Make it your mission to do your best work. Doing so demonstrates maturity and diminishes any potential tension between you and your boss. If you are producing quality work, it will get noticed by others in the company. If you and your boss have a personality conflict, do not allow your personal feelings to affect your performance. Focus on staying positive and productive, consistently being of value to your company. Even when coworkers bad-mouth the boss, do not give in to the temptation to participate. If you have a less-than-perfect boss, use the experience as a time to learn what not to do when you become a boss.

Sometime in your career, you may experience an abusive boss. The abusive boss is one who is constantly belittling or intimidating his or her employees. Abusive bosses generally behave this way because they have low self-esteem and may be feeling insecure in their position. They will utilize coercive power to make themselves feel better. There are several ways to deal with an abusive boss. If the abuse is tolerable, do your best to work with the situation. Do not speak poorly of your boss in public. If the situation becomes intolerable and is negatively affecting your performance, seek confidential advice from someone in the human resource department. Keep documentation of facts to present at this meeting. Your human resource representative may begin observing, investigating, documenting, and taking corrective action or providing needed management training to your boss if necessary. Be factual in reporting inappropriate incidents. Human resource managers want facts, not emotions. Although it may be tempting, do not go to your boss's boss. Doing so implies secrecy and distrust. Finally, if it looks as if your boss's behavior and work situation is not going to improve, begin quietly searching for another job in a different department within the company or at a new company. As an employee, you have rights. If your boss ever discriminates against you or harasses you, document and report the behavior immediately. Your boss should not make you perform functions that do not reasonably support those identified in your job description. Some bosses ask employees to run personal errands or perform duties not appropriate for the job. If you are asked to perform unreasonable functions, politely decline.

Just as we cannot choose coworkers, we can't choose our bosses. Regardless of what type of boss you have, consistently give your personal best.

Colleagues

Having friends at work is nice. Unfortunately, when workplace friendships go awry, it affects your job. Be cautious about close friendships developed at work. A close friend is someone whom you trust and who knows your strengths and weaknesses. You should be able to trust coworkers, but they should not know everything about you. While it is important to be friendly to everyone at work, there will be some with whom you want to develop a friendship outside of work, but be cautious. If there is a misunderstanding either at work or away from work, the relationship can go sour and affect both areas. Should one of you get promoted and suddenly become the boss of the other, it may create an awkward situation for both parties. Even if both you and your friend can get beyond this issue, others at work may feel like outsiders or feel you are playing favorites with your friend. Socializing outside the workplace with only friends from work places you at risk of getting too absorbed in work issues. The one common thread that binds your friendship is work; therefore, it is work that you will most likely discuss when you are away from the workplace. This can be unhealthy and could potentially create a conflict of interest in work-related decisions or cause a breach of confidentiality.

Others within the Organization

The topic of friendships in the workplace extends to those throughout the organization. Increase your professional network by meeting others within your company. As discussed in Chapter 5, when we increase our connection power, we gain additional knowledge and contacts to assist us in performing our jobs and perhaps earning future promotions. Keep conversations positive and respectful when interacting with others in the organization. Even if another individual steers the conversation in a negative direction, respond with a positive comment. Defend coworkers when another employee is talking negatively about them, and do not contribute to gossip and rumors.

| Topic Situation |

TOPIC RESPONSE

Are there other ways Julian could have appropriately responded or handled the situation?

One day, Evan and his friend, Julian, are sitting in the break room when Jolene, the department's unhappy coworker, walks in. "I just can't stand Rick!" declares Jolene. "That's too bad. Rick is a friend of mine," responds Julian. Jolene stands there red-faced, turns around, and leaves the room. Evan tells Julian that he never knew Julian and Rick were close. "Well," responds Julian, "we're not personal friends, but we all work together." Julian goes on to explain that it is easy to eliminate negative conversation when you immediately communicate that you will not tolerate bad-mouthing others. Evan thinks that this is pretty good advice.

Corporate culture is the company's personality reflected through its employees' behavior. The company's culture is its shared values and beliefs. **Employee morale** is the attitude employees have toward the company. A company's corporate culture has an enormous impact on employee morale. Morale can be positive or negative, which affects attitude, satisfaction, and employee outlook. If employee morale is positive, the employees are happier and more productive. Positive morale has a positive impact on workplace relationships. In a positive corporate culture, employers try to boost morale by recognizing and rewarding employees for good work, making the workplace comfortable, and showing trust.

Exercise 6.1

What can you do to increase employee morale in your workplace? What workplace relationships contribute to poor employee morale?

_____

_____

_____

When Relationships Turn Negative

Conflict occurs in the best of relationships. Due to a conflict or misunderstanding, a relationship may turn negative. Unfortunately, this can happen at work. Sometimes you have no idea what you have done wrong. In other cases, you may be the one who wants to end the relationship. You do not have to like everyone at work, but no one should know. Behave professionally by showing everyone respect, including your adversaries.

If you are part of a negative workplace relationship, take the following steps in dealing with the situation:

- If you harmed the other person (intentionally or unintentionally), apologize immediately.
- If the other person accepts your apology, demonstrate your regret by changing your behavior.
- If the other person does not accept your apology and your apology was sincere, continue demonstrating your regret by your improved behavior.
- If you lose the relationship, do not hold a grudge. Continue being polite, respectful, and professional to the offended coworker.
- If the offended coworker acts rude or inappropriately (either directly or indirectly), do not retaliate by returning the poor behavior. Respond in a professional manner.
- If the rude and inappropriate behavior negatively affects your performance or is hostile or harassing, document the situation and inform your boss if necessary.

The toughest step when addressing a broken relationship is when the other individual does not accept your apology. Most of us have grown up believing that we must like everyone and that everyone must like us. Because of human

Topic Situation

Josh had been one of Monique's favorite coworkers since Monique's first day at work. They took breaks together and at least once a week went out for lunch. One day Monique was working on a project with a short deadline. Josh invited Monique out to lunch and Monique politely declined, explaining why. The next day, Monique asked Josh to lunch and Josh gave her a funny look and turned away. "Josh, what's wrong?" she asked. Josh shook his head and left the room. Monique left Josh alone for a few days, hoping whatever was bothering him would blow over and things would be better. After a week of Josh ignoring Monique, things only got worse. Monique decided to try one last time to save the relationship. As Monique approached Josh, she said, "Josh, I'm sorry for whatever I've done to upset you, and I'd like for us to talk about it." Unfortunately, Josh again gave her a hollow look and walked away.

nature, this simply is not possible. We cannot be friends with everyone at work. People's feelings get hurt, and some find it hard to forgive. Behavior that is not respectful and professional interferes with performance. Your focus at work should be, first and foremost, performing your job in a quality manner. The company is paying you to perform. Therefore, if a sour relationship begins to affect your performance, do not ignore the issue. Ask yourself if your behavior is contributing to the unresolved conflict. If it is, change your behavior immediately. If your wronged coworker is upset or hurt, it is common for the coworker to begin bad-mouthing you. Do not retaliate by speaking negatively about your coworker. This only makes both of you look petty and immature. Document the facts of the incident and be mature. If the bad behavior continues for a reasonable period of time and negatively affects your performance, it is time to seek assistance from either your boss or the human resource department. This is when documentation is necessary.

When you contact your immediate supervisor for an intervention, explain the situation in a factual and unemotional manner, making sure to communicate how the situation is negatively affecting the workplace and productivity. Provide specific examples of the offensive behavior and share your documentation, including any witnesses. Do not approach your supervisor with the intent of getting the other individual in trouble. Your objective is to secure your boss's assistance in creating a mutually respectful and professional working relationship with your coworker. The boss may call you and the coworker into the boss's office to discuss the situation. Do not become emotional during this meeting. Your objective is to come to an agreement on behaving respectfully and professionally at work.

Dating at Work

A sticky but common workplace issue is that of dating other employees, supervisors, vendors, or customers of your place of employment. Because we spend so much time at work, it is natural for coworkers to seek companionship in the workplace. Although a company cannot prevent you from dating coworkers who are in your immediate work area, many companies discourage the practice. Some companies go as far as having employees who are romantically involved sign statements releasing the company from any liability should the relationship turn sour. It is inappropriate for you to date coworkers and bosses, or, if you are a supervisor, for you to date your employees. Doing so exposes you, your romantic interest, and your company to potential sexual harassment charges, which is addressed in greater detail in Chapter 12. Your romantic actions will most likely negatively affect your entire department and make everyone uncomfortable.

If you date customers, vendors, or suppliers of your company, use caution. When dating either customers or vendors, ask yourself how the changed relationship can potentially affect your job. You are representing your company 24/7. Therefore, do not share confidential information or speak poorly of your colleagues or employer. Be careful to not put yourself in a situation in which you could be accused of a conflict of interest. It is best to keep your romantic life separate from that of the workplace.

Think About It

Would you secretly date someone at work? What would be the risk? Potential harm?

Socializing

Work-related social activities such as company picnics, potlucks, and birthday celebrations are common. Though some individuals enjoy attending these social functions, others do not. You do not have to attend a work-related social function outside of your required work hours. However, it is often considered rude if you do not attend social functions that occur at the workplace during work hours. When you are working on a deadline and simply cannot spend the time to visit, briefly stop by the function and apologize to whoever is hosting the event. If you are invited to attend a work-related social function and are planning on attending, bring an item if one is requested. It is considered impolite to show up to a potluck empty-handed. It is also considered rude to take home a plate of leftovers unless they are offered. Alcohol should not be served on work premises.

Attendance at work-related social events occurring outside the work site is considered optional but provides an opportunity to network. These events include company-sponsored activities or private social functions hosted by coworkers. As discussed in Chapter 4, if the invitation requires an RSVP, send your reservation or regrets in a timely manner. Should you choose not to attend an off-site activity, thank whoever invited you in order to maintain a positive work relationship. If you do attend an off-site social function, check to see if the host has requested for guests to bring something. When the function is held at someone's home, a host or hostess gift is also appropriate. Personally thank the host for the invitation when both arriving and when leaving. If alcohol is served at an off-site work-related function, use caution if you choose to consume. It is best to not consume at all, but if you consume, limit yourself to one drink.

Shared Work Areas

In an effort to facilitate teamwork and fully utilize workspace, many work sites utilize cubicles. Although cubicles open up a work area, respect the privacy of each shared workspace as if it were an individual office. When working or conducting business in a shared work area, avoid making loud noises, smells, or distractions that may interrupt or annoy others. Speak quietly. Imagine that each cubicle has a door. Just as you would not walk into someone's office unannounced or without knocking, avoid walking into someone's cubicle without permission. Stand at the entrance of the cubicle and quietly knock or say, "Excuse me." Wait for the individual to invite you into his or her work area. If you are conducting business and the discussion is lengthy, utilize a nearby conference/meeting room so as to not disrupt others. Respect the privacy of others' workspace. Do not take or use items from someone's desk without permission.

Breaks and the Break Room

It is a common practice for offices to make coffee available. In most cases, the company does not pay for this benefit. The coffee, snacks, and other break room supplies are typically provided by employees who use them. If you routinely drink coffee or eat the office snacks, contribute funds to help pay for these items. The same goes for office treats such as donuts, cookies, or birthday cakes. If you partake, offer to share the cost or take your turn bringing treats one day. Many offices also have a refrigerator available for employees to store meals. Do not help yourself to food stored in a community refrigerator. If you store your food in this area, label your food items and throw out unused or spoiled food at the end of each work week. Finally, clean up after yourself. If you use a coffee cup, wash it and put it away when you are finished. Throw away your trash and leave the break room clean for the next person.

Miscellaneous Workplace Issues

Although it is tempting to sell fundraising items at work, the practice is questionable. Some companies have policies prohibiting it. If the practice is acceptable, do not pester people or make them feel guilty if they decline to make a purchase.

It is common and acceptable to give a gift to a friend commemorating special days such as birthdays and holidays. However, you do not have to give gifts to anyone at work. If you are a gift giver, do so discreetly so as to not offend others who do not receive gifts from you. When you advance into a management position, do not give a gift to just one employee. Managers who choose to give gifts must give gifts to all of their employees and treat everyone equally.

It is also common for employees to pitch in and purchase a group gift for special days such as Boss's Day, Administrative Assistants' Day, and retirements. While it is not mandatory to contribute to these gifts, it is generally expected that you contribute. If you strongly object or cannot afford to participate, politely decline without attaching a negative comment. If price is the issue, contribute whatever amount you deem reasonable and explain that you are on a budget. If you are the receiver of a gift, verbally thank the gift giver immediately and follow up with a handwritten thank-you note.

Exercise 6.2

What are other workplace issues that have a potential to cause conflict?

_____

_____

_____

Good employees take ownership of common work areas and practice common courtesy with such issues as refilling a coffee pot when it is empty and/or filling the copy machine when it is low on paper. When a piece of office machinery is broken or a copy machine is jammed, do not leave the problem for someone else to solve. Take responsibility and solve the problem yourself. If you are unable to solve the problem, alert someone who knows how to fix the problem.

MyStudentSuccessLab	Please visit **MyStudentSuccessLab**: Anderson|Bolt, Professionalism Skills for Workplace Success, 4/e for additional activities, resources, and outcomes assessments.

Workplace Dos and Don'ts

Do take responsibility for your performance and success at work	*Don't* wait for someone to tell you what to do
Do display consistent, professional behavior	*Don't* behave appropriately only when the boss is around
Do make your boss look good	*Don't* speak poorly of your boss
Do create positive relationships with coworkers	*Don't* make workplace friendships your primary friendships away from work
Do practice business etiquette at work-related social functions	*Don't* ignore the importance of behaving professionally at work-related social functions

Concept Review and Application

You are a Successful Student if you:

- Summarize why the concept of accountability is critical for today's workplace and provide examples of how to improve personal accountability

- Describe appropriate and inappropriate behaviors to exhibit with bosses, coworkers, vendors, and customers

- Describe how to respond when a workplace relationship turns negative
- Identify basic workplace expectations, including situations such as social functions and gift giving

Summary of Key Concepts

- Take responsibility for the job you perform by being accountable for your actions.
- Keep workplace friendships positive, but be cautious that these relationships are not your only friendships away from the workplace.
- If a workplace relationship turns negative, remain professional and respectful.
- Refrain from dating anyone at work.
- Practice good etiquette at social functions that occur within the office.

Self-Quiz MATCHING KEY TERMS

Match the key term to the definition using the identifying number.

Key Terms	Answer	Definitions
Accountability		1. The attitude employees have toward a company
Corporate culture		2. Pushing power and decision making to the individuals who are closest to the customer
Employee morale		3. The company's personality reflected through its employees' behavior
Empowerment		4. Accepting power that is being given to you
Responsibility		5. Reporting back to whoever empowered you

Think Like a Boss

1. How can you get employees excited about assuming additional responsibilities?
2. If you were to notice employee morale dropping in your department, how would you respond?
3. How would you handle two employees whose friendship had turned negative?
4. You never give your employees gifts, but one of your employees always gives you gifts for holidays, birthdays, and Boss's Day. Is it wrong for you to accept these gifts?

Activities

Activity 6.1

You have a meeting you are to attend Wednesday morning. On Tuesday evening, you have a family emergency. What should you do? Explain your answer.

_____

_____

Activity 6.2

Your boss bad-mouths or belittles coworkers. You do not like it, and you wonder what he or she says about you when you are not around. What should you do?

_____

_____

Activity 6.3

The company that services your office equipment has hired a new salesperson. This person does not wear a wedding ring and flirts with you. If you go out on a date with this person, what are three potential work-related problems that could occur?

_____

_____

_____

Activity 6.4

Identify tasks required to complete major goals in your life plan, and then list how you will be accountable to yourself to accomplish these goals. There should be several accountability measures for each life plan goal. Be specific in your answers.

Life Plan Goal	Task Required	Accountability Measures

7

Quality Organizations and Service

performance • profits • customers

After studying these topics, you will benefit by:

- Knowing how organizational structures and functions influence quality and customer satisfaction
- Naming the key elements and purpose of a company strategy
- Illustrating and interpreting an organizational chart
- Defining quality and its importance in business
- Researching methods of increasing one's creativity and innovation
- Explaining how customer service affects performance and profits
- Describing how best to handle a difficult customer

HOW DO YOU RATE?

Customer service in a quality-focused organization	True	False
1. Providing excellent service to coworkers is not as important as providing excellent service to customers.	☐	☐
2. A company's mission statement and strategy are important only to company executives.	☐	☐
3. Entry-level employees do not have a big influence on product quality.	☐	☐
4. Creativity and innovation have little influence on customer service.	☐	☐
5. When it comes to customer service, the customer is always right.	☐	☐

▶ If you answered "true" to two or more of these questions, use this as an opportunity to improve your understanding of customer service and its influence on a company's profits.

Productivity in a Quality-Focused Workplace

A quality organization exists when all employees share the same vision. Although most employees know it is important to provide excellent customer service, few understand what makes a quality organization and why customer satisfaction is critical to a company's success. Both for-profit and not-for-profit organizations need to make a profit to stay in business. A key element of profitability is productivity. Each employee is hired to perform a job that contributes to the company's success. When an employee understands the basics of how a business is structured and how it operates, the employee is better able to perform his or her job, identify creative methods to better compete and/or increase productivity, and ultimately contribute to an organization's success by providing quality products and exemplary customer service. Part of your workplace success will be based on your understanding of the business, the way it is organized, and its overall purpose.

One measure of a company's success is when it achieves its mission while making a profit. When you secure a job, you have a responsibility to the company to be productive. **Productivity** is when you are performing a function that adds value to the company. When you are productive, you are assisting the company in achieving its mission and contributing to its success. There are several ways you can add value to a company. First and foremost, behave ethically. Make ethical choices that serve the best interests of the company. Take care of company resources that have been entrusted to you, including producing quality products and eliminating waste. As you learned in Chapter 1, your attitude contributes to a company's success by influencing the attitudes of those around you, which affects productivity. Maintain a positive attitude, strive to create positive workplace relationships, and be open to learning new skills.

Each company needs a strategy. A company's **strategy** outlines major goals and objectives that serve as its road map for success. A **strategic plan** is a formal document that identifies how the company will secure, organize, utilize, and monitor its resources. Together, the company's mission, vision, and values statements constitute the organization's **directional statements**. Directional statements create the foundation for why the company exists and how it will operate.

A company's **mission statement** is its statement of purpose. It identifies why each employee comes to work. For example, part of a college's mission statement is to contribute to student success. This means not only providing a solid education, but also preparing students to succeed while in school, on the job, or with their continuing education. If all college employees know that their ultimate purpose is to contribute to student success, each activity they perform on the job should contribute to student success. Either prior to starting your job or within your first week at work, secure a copy of your company's mission statement. If possible, memorize it. A company's mission statement is typically included in the employee handbook and/or provided to you during the employee orientation. If this information is not provided, check the company's website or ask your supervisor or the human resource department for a copy.

Once a company has identified why it exists through its mission statement, it determines where it wants to be in the future. This is called a **vision statement**. A company's vision statement is its viable view of the future. For example, a college may want to be the top-ranked college in the nation. This goal will take work, but it is achievable. In addition to the company's mission and vision statements, each company should have a values statement. The **values statement** defines what is important to and what the priorities are for the company. Common company values include providing a healthy return to investors, taking care of the environment, taking care of its employees, and keeping customers satisfied. An important element of the company values statement is the company's code of conduct or ethics statement. These statements discuss the importance of behaving ethically in all areas of business. Companies are accountable to various **stakeholders**, who are entities that have a direct interest in a company. Key stakeholders include investors, employees, customers, and the community at large. Factors within the business environment change, so companies should constantly monitor changes and be proactive in meeting the needs of their various stakeholders.

Topic Situation

Juan's company was updating its strategic plan and asking for volunteers to sit on various planning subcommittees. Juan wondered if new employees were able to sit on these committees. Juan decided to ask a coworker. The coworker encouraged Juan to participate and said it would be a great way for him to learn more about the company, network with people throughout the organization, and help make positive changes for the company's overall success. Juan's coworker then showed him the back side of his name badge. On it was printed the company's mission statement. "This," Juan's coworker proudly said, "is why we come to work every day." Juan signed up for a committee that afternoon.

TOPIC RESPONSE

What additional advantages will Juan gain if he joins the committee? How can Juan prepare for this experience?

Company resources include financial (fiscal), human (employees), and capital (long-term investments) resources. Many companies provide summaries or overviews of the strategic plan to all employees to keep the organization focused. This information is often found on the company website. Just as you created a personal plan earlier in this text, each area of the company will have smaller plans with stated goals and objectives that identify how their respective areas will assist in achieving the company's strategy. As defined in Chapter 1, a **goal** is a target. In a business environment, a goal describes what needs to be achieved to contribute to the success of the company strategy (respective of the position or department setting the goal). An **objective** identifies specific activities that support a goal. Objectives are sometimes referred to as *short-term goals*. An objective must be measurable and have a deadline. Each area of the company will utilize its respective resources and have its performance monitored based on the company strategy, goals, and objectives.

Exercise 7.1

Assume it is your first day in your target career job. Write one goal and two objectives for your position.

Goal:

Objective 1:

Objective 2:

Lines of Authority

Functions and resources within a business are organized according to the company's mission and strategy. The way a company is organized is called its **organizational structure**. The graphic visual display of this structure is called the company **organizational chart**. This chart not only identifies key functions within the company, but also shows the formal lines of authority for employees. These formal lines of authority are also referred to as the **chain of command**. The chain of command identifies who reports to whom within the company. Respect and follow the formal lines of authority within your company. For example, based on the organizational chart in Figure 7.1, it is inappropriate for the accounts payable supervisor to directly make a request to the marketing vice president without prior approval from the accounting director.

Topic Situation

One day Lucia was left alone in the department while everyone was away at an important meeting. A tall man walked through the door and asked to see the boss. Lucia explained that the boss was currently out. The gentleman asked her how she felt about the company, including how long she had worked for the company and what could be done to improve the company. Lucia found the questions strange but interesting. She shared ideas on how to improve the company in a polite, positive, and constructive manner. The gentleman thanked Lucia for the input and left. Two weeks later, Lucia's boss returned from a meeting and told her that the CEO of the company had shared Lucia's ideas on how to improve the company with key executives. It was then that Lucia realized that the earlier conversation was with the CEO of the company.

TOPIC RESPONSE

Why should you know a company's organizational structure and identities of key executives?

Each company has a leader (see Figure 7.2). The leader is typically called the **president or chief executive officer (CEO)**. The president or CEO implements company strategy and reports to the company **board of directors**. This group of individuals is responsible for developing the company's overall strategy and major policies. The company's board of directors is elected by shareholders or investors. Smaller companies may not have such a formal structure and/or

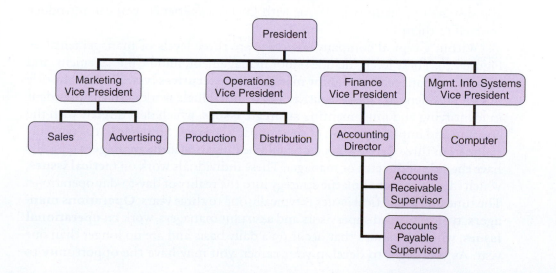

Figure 7.1

Organizational Chart

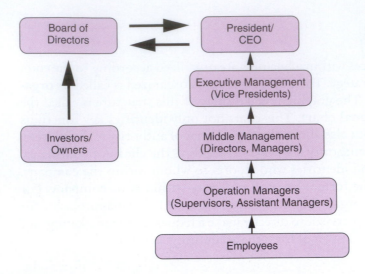

Corporate Structure

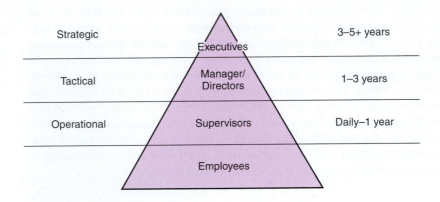

Management Levels

titles, but each business has investors/owners and a leader (president). Become familiar with your leaders' names and titles and the formal lines of authority. This will allow you to determine whom you can go to if there is a question or problem. When possible, view updated photos of these individuals so that, if the appropriate opportunity arises, as with Lucia's experience, you can introduce yourself to them.

Within a typical company structure are three levels of management (see Figure 7.3). These levels include senior management, middle management, and operations management. **Senior managers or executives** typically have the title of vice president or chief executive. These individuals work with the president in identifying and implementing **strategic issues**, which deal directly with the creation and implementation of the company strategy. Strategic issues typically range from three to five years, but could be longer. **Middle managers** typically have the title of director or manager. These individuals work on **tactical issues**, which identify how to link the strategy into the reality of day-to-day operations. The time line for tactical issues is typically one to three years. **Operations managers**, typically called supervisors and assistant managers, work on **operational issues**, which are issues that occur on a daily basis and are no longer than one year. As you begin to develop your career, you may have the opportunity to advance into a management position. Your first step into a management position will be that of a supervisor. **Supervisors** are first-line managers. Although supervisors primarily concern themselves with operational issues, successful employees and supervisors understand the bigger picture of the company's overall tactics and strategy.

As displayed in a typical organizational chart, companies are usually arranged by major functions. These major functions are frequently referred to as **divisions**. Within these divisions are departments. The **departments** are subdivisions responsible for carrying out specific tasks of their division. Several major functions (divisions) are necessary in business. These include finance and accounting, human resources, operations, information systems, marketing, and legal counsel.

The **finance and accounting** area is responsible for the securing, distribution, and growth of the company's financial assets. Any invoices, incoming cash, or checks must be recorded through the accounting department. The accounting department will work with the human resource area on payroll issues regarding your paycheck. Just as you have a personal budget (refer to Chapter 2), companies utilize several types of budgets. A **capital budget** is used for long-term investments, including land and large pieces of equipment. An **operational budget** is used for short-term items including payroll and the day-to-day costs associated with running a business. Because the primary purpose of every company is to make a profit, you as an employee are accountable for how you utilize the company's financial resources.

Topic Situation

One of Max's coworkers went on vacation. Before leaving, the coworker showed Max how to order office supplies online and asked Max to order supplies if needed. Max noticed that the department was running low on a few items and decided to place an order. As Max clicked through the online catalog, he saw a lot of items that would be nice to have around the office, including a hole punch, a label maker, and a portable hard drive. Max also thought it would be nice to have a pair of new scissors and a USB flash drive. As Max was about to place the order, he was shocked to see that the total was over $1,000. Max realized that the coworker never gave Max a budget, but he knew that one existed.

TOPIC RESPONSE
What should Max do? Justify your response.

As illustrated in Max's example, employees frequently do not think before they spend the company's money. Before you spend the company's money, ask yourself a simple question: "If I owned this company, would I spend my money on this item?" Answering that simple question makes you more accountable for your actions and makes you think like a business owner. You most likely will not spend your money on frivolous, unnecessary items and will pay more attention to not only adhering to a budget, but also identifying ways to save money.

The **human resource** area is responsible for recruiting, hiring, training, evaluating, compensating, promoting, and terminating employees. This area deals with the employee (people) side of business. The first contact you will have with this function will be when you apply and interview for a job. You will also interact with the human resource area concerning any issue relating to company policy, complaints and grievances, and your terms of employment. The human resource function is discussed in greater detail in Chapter 8.

The **operations** function deals with the production and distribution of the company's product. It is the core of the business. Even if your position does not directly contribute to the production of a company product, your job supports individuals who produce, sell, and distribute the product.

The **information systems (IS)** area deals with the electronic management of information within the organization. This function is responsible for ensuring that the company appropriately utilizes its computer/technology resources. As an employee, you assist and support the information systems function by utilizing company technology only for work-related business. Responsible employees know and practice computer basics such as routinely backing up files, emptying the electronic trash bin, and conducting routine virus checks.

Marketing is responsible for creating, pricing, selling, distributing, and promoting the company's product. The company's **legal counsel** handles all legal matters relating to the business. Check with the company's legal department prior to engaging in a contract on behalf of the company. Large companies may have separate divisions for each of these functions, whereas smaller companies combine several functions into a single division, department, or position. As stated earlier, not all companies have the formal departments, titles, or organizational charts described in this chapter. Most small businesses will not have such formal structures. However, to be successful, they will have someone who performs these important functions for the business.

Quality and the Company

All companies offer or sell a **product**. A product is what is produced by a company. Products come in the form of goods and services. A **good** is a tangible item, something that you can physically see or touch. Appliances, toys, or equipment are examples of tangible products. A **service** is an intangible product. Examples of services include haircuts, banking, golf courses, and medical services. Customers demand quality products. If customers do not perceive that they have received a quality product, they will not make a repeat purchase.

A quality product, excellent service, and innovation are what will persuade customers to make a purchase from a specific company. A company cannot experience long-term success if one of these elements is missing. **Quality** is a predetermined standard that defines how a product is to be provided. Customers measure quality by comparing your company's product to similar products. They also measure quality by how satisfied they are after using a product. Successful companies include performance monitors in their strategies. Performance monitors identify how quality will be measured, assess areas of quality and service that are performed well, and also identify areas that need improvement. Monitors, or standards, may include defect rates, expenses, or sales quotas.

Another example of measuring quality is an employee evaluation. The evaluation can identify how an employee's performance contributes to customer quality. Quality-focused evaluation criteria address areas such as response time, attitude toward customers, and the proper use of resources. An expanded discussion on employee evaluations is presented in Chapter 8.

Companies want to build brand loyalty with customers and do not want customers to substitute their product with a competitor's product. Customers are loyal to a company and its products when they receive **value**, which means customers believe they received a good deal for the price they paid. Providing a quality product at a reasonable price (value) requires an ongoing company-wide effort. A successful company must consistently make a profit. Profit is revenue (money coming in from sales) minus expenses (the costs involved in running the business). Assist your company in achieving this goal by monitoring and identifying methods to decrease expenses and increase sales. Be aware of unnecessary expenses you incur at work and make every effort to eliminate waste. Find ways to be involved with and take responsibility for better knowing customer needs. As profits increase, the company will have the ability to grow. This growth may include expanding into a larger space, adding more sites, offering additional services and/or goods, increasing pay and benefits, and/or hiring more employees. For an employee, this growth could result in a raise or promotion. Companies that cannot provide a quality product at a reasonable price will not experience long-term success. Employees play an important role in helping a company achieve this goal. This is why every employee must take responsibility for ensuring quality and eliminating waste in his or her specific job.

Talk It Out

Identify common money wasters in the workplace.

Exercise 7.2

As an employee, list ways you can take personal responsibility for quality at work.

_____

_____

_____

Creativity and Innovation

As the global economy becomes increasingly competitive, it is necessary for individuals to enhance their creativity. **Creativity** is coming up with a new and unique item, service, or system. Creativity is achieved when an individual identifies a new use or application for an object or situation. Creativity can occur only when an individual is not restrained by traditional thinking. A creative person will always ask, "What if?"

Consider one of America's greatest inventors, Thomas Edison. He developed the phonograph when he was trying to improve the efficiency of the telegraph. Through the creativity of others, the phonograph evolved into a record player, which evolved into a portable music device, which evolved into a mobile communication device that includes a phone, Internet access, and nearly unlimited music storage. Imagine how life would be without mobile devices if Edison had not developed the phonograph.

Employees need to work on enhancing their workplace creativity. This is done by looking at situations differently and identifying how to improve current systems and/or methods in an effort to improve quality, reduce costs, and increase customer satisfaction. Doing so opens doors for new products and increased efficiencies. Although creativity is important in the workplace, it is not useful if the new ideas are not acted on. **Innovation** is the process of turning a creative idea into reality. Continually identify new uses and applications for items and/or situations and then act on those new ideas. Work on improving your creative and innovative skills in an effort to improve and/or contribute to your company's success.

Web Search

Search online to find and take a quiz to test your customer service knowledge and skills.

Excellent Customer Service Defined

An important business concept relating to workplace quality is customer service. **Customer service** is the treatment an employee provides the customer. Companies cannot survive without customers; therefore, you need to know who your customers are and how to treat them.

A company has internal and external customers. **Internal customers** are employees within a company. **External customers** are individuals outside of the company. External customers include the individuals or businesses that purchase a company's product, vendors, and investors. You may have a job where you do not directly interact with the company's external customers. If such is the case, you still have an obligation to serve and treat internal customers as well as the company expects its employees to treat external customers.

Talk It Out

Discuss the difference between a service and customer service.

Exercise 7.3

As a customer, how do you expect to be treated? How do you expect your coworkers to treat you?

_____

_____

_____

Maintaining customer satisfaction is one of the best ways to sell a product. Satisfied customers will not only make repeat purchases, but will also encourage others to purchase your product. Similarly, unhappy customers also tell others about their bad experiences. Given the increased use of social media to share customer experiences, both favorable and unfavorable experiences can quickly go viral over the Internet. No business wants customers bad-mouthing the company or its product.

Employees need to be competent and dependable. A competent employee knows the product(s) his or her company offers and is able to answer common customer questions. Customers expect employees to assist them by providing correct information about a product. Dependable employees are reliable and take responsibility in assisting customers. Do not pretend to know something when you do not know the answer to a customer's question. You will gain customer respect when you admit that you do not know all the answers, but can find someone who can assist the customer. If there is a situation that requires you to seek assistance from another employee when helping a customer, do not just hand the customer over to the other employee without explaining the situation. Whenever possible, stay with the customer and learn from your coworker so that you will know the answer the next time someone asks.

Strive to provide customers with personal attention and try to anticipate customer needs. Some customers like to be left alone to shop for a product but want you near if questions arise. Other customers would like you to guide them step by step when purchasing a product. When you see a customer, make every effort to acknowledge him or her as soon as possible with a greeting and an offer of assistance. Use the customer's name if you know it to create a more personal and friendly atmosphere. Customers need to be treated differently according to their needs. Many companies provide customer service and sales training specific to their industry. If provided the opportunity, take advantage of such training.

Customers also expect a welcoming, convenient, and safe environment. This includes the appearance of the building, as well as the appearance of employees. As soon as a customer comes in contact with a business, an opinion is formed about that business. There is only one first impression, so it must be positive. The appearance of the building and/or employees can be the reason a customer visits your company in the first place. Whether you work directly or indirectly with customers, you have an obligation to keep your workplace clean and immediately address potential safety hazards. If there is trash on the floor or a spill of water, clean it up as soon as possible. Take responsibility for keeping your workplace clean and safe. Your attitude, language, and attire contribute to the image of your company. Display loyalty to your company by consistently supporting the company and its mission through your actions. Do your job and do it well. Respect company policies, your coworkers, and customers. Make every effort to promote the company and its products. Understand your company, including its strategy and business structure. You are a walking billboard for your company. Your behavior both at and away from work represents the company. Do not speak poorly of your company, your coworkers, or the company's product either at or away from the workplace.

The Impact of Customer Service

Excellent customer service is the gateway to a company and the biggest reason customers return. Satisfied customers keep returning when you make them feel valued and provide them a high-quality product at a reasonable price. Happy customers also tell others about your business. With so many choices and increased competition, the same product can be purchased from many different businesses. Your goal is to build a long-term relationship with customers that will make them loyal to you and your business. Customer loyalty is another important element that contributes to the success of a company. If a customer perceives he or she has received value and a quality product, he or she will display loyalty to your company by making a repeat purchase. Companies want to build brand loyalty with customers and want customers to not substitute their product with a competing product. Customers will be loyal to a company and its products when quality products are consistently provided.

This is the reason so many businesses now keep electronic records of their customers. The more information businesses maintain on their customers, the more they are able to personalize service. With the increased use of technology, many companies maintain databases of information, including past purchases, preferred payment methods, birthdays, special interests, and return/exchange practices. Companies utilize this information to establish a more personal relationship with the customer and anticipate future needs. By connecting with customers through e-mail and various social networking sites, it is common to notify customers of upcoming sales of frequently purchased products, send notification of upcoming events, or disperse discounts and coupons. Customer information and records are confidential and should be used only for business purposes.

Talk It Out

How does personal appearance influence a customer's trust and perception of you?

TOPIC RESPONSE

What specific actions did the pizza shop take to display excellent customer service to Kiana?

When Kiana moved to a new neighborhood and wanted to order pizza, she found there were many different pizza parlors. The way to determine which one to try was to ask others for a recommendation or randomly try each one. Because Kiana was new to the area and did not know many people, she called one of the pizza shops. When Kiana called, the employee was friendly and sounded happy to assist with the order. Kiana automatically formed a good impression of the pizza shop. When she picked up her order, the inside of the shop was clean, as were the employees. The pizza also tasted good. The next time Kiana wanted a pizza, she called the same place.

Talk It Out

At work, how should you handle a situation when you observe a verbal conflict between two employees that is taking place in front of a customer?

The success and profitability of a company depend on how you treat your customers. The happier customers are, the more likely they are to return. A business needs satisfied customers to not only make repeat purchases, but also tell other potential customers about their positive experience.

An important element of customer service is how customers view employee-to-employee interactions. Customers have more respect and credibility for a company when they witness positive interactions between employees. Employee conflicts should not be visible in front of customers. These matters and related discussions should be kept private and away from customers or it will reflect poorly on the entire organization.

The Difficult Customer

Customers can sometimes be difficult to deal with. Historically, companies have had the motto "The customer is always right." But, in many instances, the customer may not be right. Although the customer may be wrong, adopt the attitude that the customer is unhappy and do all you can to help the customer solve a problem. Have patience and sympathize with the customer to maintain a positive relationship.

Think About It

How do you normally respond to conflict? Does your response hinder or help you provide appropriate service to a difficult customer?

Many times, a difficult customer will be unfriendly and may even begin yelling at you. If this occurs, stay calm and do not take the customer's inappropriate behavior personally. By remaining calm, you are better able to identify the real problem and logically get the problem solved as quickly as possible in a manner that is fair to both the customer and your company.

In order to successfully resolve a difficult customer's complaint, do the following:

1. *Stay calm, let the customer talk, and listen for facts.* This may mean letting the customer vent for a few minutes. Although it is not easy to stay calm when someone is yelling at you, do not interrupt or say, "Please calm down." Doing so will only increase the anger. Pay attention, nod your head, and take notes if it helps you keep focused. As stated earlier, although the person may be yelling at you, do not take the harsh words personally.

2. *Watch body language.* Be mindful of the customer's tone of voice, eye contact, and arm movement. If a customer avoids eye contact, he or she

may be lying to you or not fully conveying his or her side of the story. Do not allow a customer to touch you, especially in a threatening manner. If you feel a difficult customer has the potential to become violent or physically abusive, immediately seek assistance.

3. *Acknowledge the customer's frustration.* Use calming phrases such as "I can understand why you are upset." Let the person know you have been listening to his or her concern by paraphrasing what you have understood the problem to be. Do not repeat everything; summarize the concern to ensure an understanding of the facts.

4. *Find a resolution.* Whenever possible, take care of the problem yourself. Sending the customer to another person may only fuel the customer's frustration. Though it is tempting to call your supervisor or a coworker, stay with the customer until you know the problem has been solved.

5. *Know company policy.* Some difficult customers are dishonest and attempt to frazzle employees with intimidation and rude behavior in an attempt to defraud the company. Know company policies and do not be ashamed to enforce them consistently. If a customer challenges a policy, calmly and politely explain the purpose of the policy.

6. *Expect conflict, but do not accept abuse.* Difficult customers are a fact of life. Although customers may occasionally yell, you do not have to take the abuse. If a customer shows aggressiveness or is cursing, politely state that you cannot help solve the problem until the customer is able to treat you in a respectable manner. If the customer continues the inappropriate behavior, immediately call a supervisor.

> **Talk It Out**
>
> If a customer is angry with a raised voice, how should you respond to that customer?

MyStudentSuccessLab Please visit **MyStudentSuccessLab**: Anderson|Bolt, Professionalism Skills for Workplace Success, 4/e for additional activities, resources, and outcomes assessments.

Workplace Dos and Don'ts

Do read the company mission statement so you remember why the company pays you to come to work each day	*Don't* ignore the company's directional statements and their application to your job
Do know your internal and external customers as well as what role you play in ensuring quality and contributing to your company's success	*Don't* assume that your only customer is outside of the company and that you have no influence on the company's overall success
Do take responsibility for producing and/or providing quality	*Don't* ignore quality by allowing wasted materials
Do be a role model for other employees by eliminating waste and increasing productivity	*Don't* allow bad attitudes to affect your performance

Do display competence by knowing your company products and policies	*Don't* lie to customers and make up information you don't know regarding company products and policies
Do remain calm when dealing with a difficult customer and seek assistance immediately if a customer becomes abusive	*Don't* tolerate foul language or violence

Concept Review and Application

You are a Successful Student if you:

- Define and explain why the concepts of quality, productivity, and customer service are critical in today's economy
- Create, label, and explain the elements of an organizational chart for a traditional company.
- Outline specific steps you will take to increase your creativity

Summary of Key Concepts

- Directional statements include the company's mission, vision, and values statements.
- The company's strategic plan identifies how a company will secure, utilize, and monitor resources for success.
- A company's organizational chart is a graphic display of the major functions and formal lines of authority within an organization.
- Major functions (divisions) that are necessary within a business include finance and accounting, human resource management, operations, information systems, marketing, and legal counsel.
- Excellent service, quality, and innovation are what will persuade customers to purchase a company's product or service.
- Employees should work on improving their creative and innovative skills in an effort to contribute to a company's success.
- Customers can be internal customers (other employees) or external customers (individuals outside of your company). A successful company has concern for both internal and external customers.

- Employees who provide excellent customer service are competent, dependable, and responsive.
- The customer is not always right, but you need to adopt the attitude that if the customer is unhappy, you must do all you can to help the customer solve his or her problem.

Self-Quiz MATCHING KEY TERMS

Match the key term to the definition using the identifying number.

Key Terms	Answer	Definitions
Board of directors		1. Budget used for long-term investments
Capital budget		2. Responsible for the securing, distribution, and growth of a company's financial assets
Chain of command		3. Deals with the electronic management of information within an organization
Creativity		4. Issues that occur on a daily basis and no longer than one year
Customer service		5. Executive responsible for implementing a company's overall strategy
Departments		6. Managers that concern themselves with operational issues
Directional statements		7. Formal document that identifies how the company will secure, organize, utilize, and monitor its resources
Divisions		8. A tangible item
External customer		9. Company statement that defines what is important to a company
Finance and accounting area		10. First-line managers
Goal		11. Subdivisions responsible for carrying out specific tasks of a division
Good		12. A company's statement of purpose
Human resource area		13. A company's formal lines of authority
Information systems area		14. Company statement that determines where it wants to be in the future
Innovation		15. A predetermined standard that defines how a product is to be produced
Internal customer		16. A graphic display of a company's structure
Legal counsel		17. Statements that create the foundation for why the company exists and how it will operate
Marketing		18. Identifies specific activities that support a goal

Key Terms	Answer	Definitions
Middle managers		19. Entities that have a direct interest in a company
Mission statement		20. A customer's perception of receiving a product at a fair price
Objective		21. What is produced by a company
Operational budget		22. Coming up with a new and unique item, service, or system
Operational issues		23. Handles all legal matters relating to a business
Operations		24. Responsible for recruiting, hiring, training, and other issues related directly with employees
Operations managers		25. Major functions of a company
Organizational chart		26. Issues that deal with aligning the strategy with day-to-day issues
Organizational structure		27. Individuals responsible for identifying and implementing strategic ideas
President or chief executive officer (CEO)		28. Individuals outside of the company
Product		29. The treatment an employee provides the customer
Productivity		30. A target that describes what needs to be achieved to contribute to the success of a company strategy
Quality		31. Fellow employees within a company
Senior managers or executives		32. The way a company is organized
Service		33. Budget used for short-term items
Stakeholder		34. Responsible for creating, pricing, selling, distributing, and promoting a company's product
Strategic issues		35. An intangible product
Strategic plan		36. Outlines major goals and objectives which serve as its road map for success
Strategy		37. Company issues that deal directly with the creation and implementation of a company strategy
Supervisor		38. The process of turning a creative idea into reality
Tactical issues		39. Deals with the production and distribution of a company's product
Value		40. Group of individuals responsible for developing the company's overall strategy and major policies
Values statement		41. Individuals responsible for tactical issues
Vision statement		42. Performing a function that adds value to a company

Think Like a Boss

1. You are the supervisor for a team of employees who have a high number of product defects. They also waste materials. You recognize that product defects and wasted materials affect your department's budget. You have told your team to decrease the amount of wasted materials, but your employees do not seem to care. How can you get them to increase their quality and decrease waste?

2. One of your best customers verbally abuses two of your employees every time she visits your store. Your employees have complained to you several times about this customer. What should you do?

Activities

Activity 7.1

Review the following organizational chart and answer the following questions.

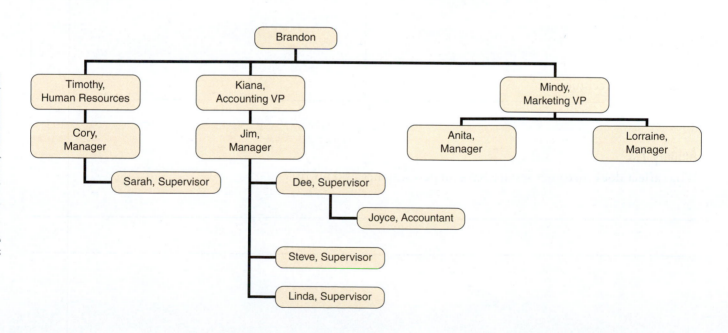

1. To whom should Linda go if there is a question regarding employee benefits?

   _____

2. Who is Joyce's immediate supervisor?

   _____

3. If Joyce's immediate supervisor is not available, from whom should she seek assistance?

   _____

4. Who is ultimately responsible for creating, pricing, selling, distributing, and promoting the company's product?

   _____

5. What is Brandon's title?

   _____

Activity 7.2

How would you measure performance in the following jobs?

Job	Performance Measures
Office Professional	
Paralegal	
Medical Biller	
Chef	
Auto Repair Technician	

Activity 7.3

What affect does customer service have on performance and profits?

_____

_____

Activity 7.4

What is the most creative thing you have done to help you achieve your career goal?

_____

_____

How can you integrate creativity in your personal understanding and commitment to improve quality, efficiency, and customer satisfaction?

_____

_____

Activity 7.5

Describe a time you received exemplary customer service. Be specific in identifying the treatment you received.

_____

_____

Human Resources and Policies

benefits • considerations • evaluation

After studying these topics, you will benefit by:

- Stating the purpose of and naming key services the human resource department provides employees
- Listing the primary elements of an employee orientation program
- Describing the purpose of an employee handbook
- Identifying the types of employee status and purpose of an introductory period
- Defining performance evaluations, their purpose, and explaining an employee's role in the process
- Providing a general overview of employee benefits
- Explaining the purpose of a union and its benefits

HOW DO YOU RATE?

Employee rights	True	False
1. All full-time employees must receive at least two weeks' paid vacation.	☐	☐
2. All full-time employees must receive company paid vision and dental benefits.	☐	☐
3. Employers do not have the legal right to change policies once they are printed in the employee handbook or listed on the employer website.	☐	☐
4. Employers cannot legally fire an employee without two weeks' notice.	☐	☐
5. Employees cannot legally quit without providing the employer two weeks' notice.	☐	☐

▶ If you answered "true" to two or more of these questions, take this opportunity to learn more about employee and employer rights. Doing so makes you a better employee and future boss.

Human Resource Department

One of the first departments you will interact with at a new job is the **human resource department (HR)**. This department is responsible for hiring, training, compensation, benefits, performance evaluations, complaints, promotions, and changes in your work status. Apart from the boss, the human resource department is an employee's primary link to the employer. The purpose of this chapter is to share common policies and resources that are accessed through the human resource department. Being informed and utilizing the resources provided by this department makes you a more productive and valued employee.

Employee Orientation

Typically within the first few days of employment, you will receive an **employee orientation**. This is the time when the company's purpose, structure, major policies, procedures, benefits, and other important matters will be explained. You may also be issued company property at this time, including an employee ID and keys.

When you studied the concept of quality customer service, you learned that understanding a new company includes knowing the major products the company provides as well as key individuals in charge. This information is reflected in the company's mission statement and organization chart. Pay attention to the names and titles of key executives, including the company president, CEO, and vice presidents. Utilizing the company's printed materials, its website, or an Internet search, view current photos of these individuals, if available, so that if you have the opportunity to meet these executives, you will know who they are.

During your first few days at work, make an effort to meet coworkers and identify potential mentors. A **mentor** is someone who can assist you in developing leadership skills, provide support, expand your professional network, and help you grow in your career. Mentors also provide you the opportunity to improve job skills and increase your potential for career advancement. A mentor will help you understand the culture of the company. The **corporate culture** includes the values, expectations, and behaviors of people at work. By understanding a company's corporate culture, you will learn about its politics, its policies, and how coworkers and executives expect you to behave on the job.

Finding a mentor can be a formal or informal process. Some companies have formal mentoring programs. In this instance, you will have a mentor who is assigned to help you succeed at work. Assigned mentors are often paid by the company to allow mentoring during work hours.

If your company does not offer a formal mentoring program, try to establish an informal mentoring relationship with someone who is willing to help you learn about your new job and career. A true mentor relationship cannot be forced because the relationship is based on trust. Be someone with whom an executive would like to invest his or her time by showing a willingness to learn and grow. You do not have to be employed to have a mentor. If you are not employed, select someone whom you consider a leader, knows your career plans, and works in your targeted industry.

Talk It Out

What are common questions employees have on their first day of work?

Employee Handbook

After a general overview of the company, you will be given a booklet or a web link to the electronic version of the employee handbook. An **employee handbook** outlines an employee's agreement with the employer regarding work conditions, policies, and benefits. Some of these policies are legally required; others address rules of conduct and/or benefits. Keep this information accessible and use it as a reference for major workplace issues. A representative of the company will review important sections of the handbook with you. Ask questions on topics that you do not fully understand. After the handbook is reviewed, you will usually be asked to sign a statement affirming that you have received the handbook, read it, and agree to its contents. This is a legal agreement. Therefore, do not sign the statement until you have completely read the handbook and fully understand its contents. Most employers will provide the employee a day or two to return the signed agreement.

Topic Situation

TOPIC RESPONSE

What should Oliver do?

During Oliver's employee orientation, Oliver received the Internet link to the employee handbook. With all new employees present, a representative from the human resource department explained the handbook. Unfortunately, Oliver was so overwhelmed with new terms, policies, and signing forms that he was not confident that any information was truly understood. After the orientation, Oliver placed the orientation paperwork in a drawer at home. About three weeks later, Oliver's friends wanted to take a mini-vacation on an upcoming holiday and invited Oliver. Oliver was not sure the company gave that day as a holiday.

Employment-at-Will and Right to Revise

Legally required in many states, a major policy statement that is usually placed at the beginning of an employee handbook is called employment-at-will. **Employment-at-will** policies do not contractually obligate employees to work for the company for a specified period. This policy applies to any employee who is not hired as a contract employee for a stated period of time. Contract employees literally have contracts outlining the terms, start dates, and end dates of employment. If you are an employee-at-will, you may quit anytime you want. By the same token, your employer may terminate your employment at any time.

An employer also has the **right to revise**, which is a statement contained in an employee handbook that provides an employer the opportunity to change or revise existing policies. You may be asked to sign separate statements affirming that you understand both the employment-at-will and right-to-revise policies if they are applicable.

There should also be policies in the employee handbook regarding equal employment opportunity and discrimination, which state that the company does not discriminate or allow unlawful harassment of any kind, including sexual harassment, a hostile workplace, or hate crimes. Harassment and discrimination are addressed in greater detail in Chapters 12 and 15, respectively.

Web Search

Find answers to questions on various labor topics by exploring the Frequently Asked Questions page on the U.S. Department of Labor website.

Employment Status

An additional section of most employee handbooks addresses employment status definitions, including the differences between a full-time employee, a part-time employee, and a temporary employee. These classifications are typically determined by the number of hours worked per week and/or the length of employment with a company.

Full-time employees work a pre-determined number of hours per week (normally 30 or 40, depending upon the industry) and are eligible for employer benefits. Unless the position is classified as exempt, employees who work more than 40 hours per week are entitled to overtime pay. **Part-time employees** have varied work hours and normally do not qualify for employer benefits.

Temporary employees are hired for a specified period of time, typically to assist with heavy work periods or to temporarily replace an employee on leave. Although that relationship still exists, more employers are hiring new employees as temporary employees to first determine if they are a good fit with the company. If the employee does a good job, the employer will offer the employee a regular full-time position. If the temporary employee is not a good fit with the company culture, the employee is simply let go. For most companies, all new employees hired enter into an introductory period. The **introductory period** (also known as a *probation* or *orientation period*), is typically the first one to three months of employment in which employers are provided time to evaluate a new hire's performance and determine if the new hire should continue as an employee. The new hire also has that period to decide if he or she wants to continue working for the employer. Near the end of this period, the new hire may be given a performance evaluation. If the employer is satisfied with performance, the new hire will become a regular, full-time employee and begin receiving benefits and/or other entitlements of a full-time employee. If the employer is not satisfied with the new hire's performance, he or she can terminate the employee without cause during the introductory period. If the performance is not yet acceptable but the employer thinks the new employee demonstrates potential, the introductory period may be extended.

Although the following information should be contained in an offer letter, it is your responsibility as a new employee to identify:

- If you are being hired as a temporary, part-time, or full-time employee
- Whether your company has an introductory period

- The length of your introductory period
- If and when you become eligible for benefits (if offered)
- Factors that will be used to evaluate your job performance

After determining the length of your introductory period, secure a copy of your job description and performance evaluation. A **job description** outlines who you report to, exempt or non-exempt status (to determine if you are entitled to overtime pay), job duties and responsibilities (i.e., why a company is paying you to come to work), and qualifications and skills needed to perform the job. A **performance evaluation** identifies how work performance will be measured. Performance evaluations contain various criteria that measure an employee's daily productivity, efficiency, and behavior. Common factors used to evaluate performance reflect the duties and responsibilities included in a job description. Additionally, your involvement in work-related activities, ongoing education, ability to work with both internal and external customers, and ability to assume new responsibilities may be reflected in the evaluation. Both your job description and your performance evaluation will assist you in becoming a better employee and may provide opportunties to advance within the company.

Performance Evaluations

Most employers provide performance evaluations immediately after completing an introductory period and once a year thereafter. Although the prospect of someone providing feedback on your performance can be a bit intimidating, use this time to obtain information on how to become a better employee. Performance evaluations provide a time not only for your supervisor to provide performance feedback, but also for you to share your desire for additional training and responsibilities. Employees should receive advance notice of an impending evaluation and performance criteria. Based on the established criteria, keep an ongoing record of your performance. This should include notes and letters from customers, coworkers, and vendors. It may also include personal documentation of events detailing work accomplishments and times you displayed excellent judgment and/or behavior. On occasion, a supervisor provides the employee a blank copy of the evaluation form and asks the employee to complete a self-assessment. Use this opportunity to provide an honest review of your performance. Do not be overly favorable or overly critical in your self-assessment. Be honest. Refer to the evidence and documentation you collected to support your assessment. After you have completed your self-assessment, make a photocopy of your document and return the original to your supervisor.

Arrive at your evaluation meeting with a positive attitude and view the evaluation process as a time for constructive feedback and career growth. Bring whatever performance documentation you have collected. During your formal evaluation, sit quietly and listen to your supervisor's assessment of your performance. If you receive a favorable evaluation, thank your supervisor for the assessment and express your desire for increased responsibilities and/or opportunities. If there is anything included in the evaluation that you do not agree with, take notes but do not interrupt your supervisor. Share your concerns only when your supervisor is finished talking or asks for feedback. Support your comments with facts and documentation. Even if you do not agree with your supervisor's response after you have presented your evidence, do not argue or challenge your supervisor during the assessment.

Think About It

Why is it important to perform your best every day at work?

At the end of each appraisal form is an area for the employee's signature. Immediately under the signature area should be a statement that the employee's signature does not constitute agreement with everything contained in the assessment, only proof that the employee received an evaluation. If you do not agree with any statement included in your evaluation and the appraisal form does not contain the preceding statement, do not sign the evaluation. If the appraisal form does include the statement, sign the evaluation but attach a written response regarding any areas you specifically do not agree with and why. Provide supporting documentation and evidence to your attached statement. Do not write an emotional response. Make your statements factual and professional and do not attack your supervisor or anyone else. Employers usually allow employees one day to provide a written response. Both your original evaluation and your written response will be forwarded to the human resource department and will become a permanent part of your personnel file. Keep photocopies of your evaluations. They serve as legal documentation of your performance. If your evaluations are favorable, the written evaluation also serves as excellent reference material for future employers.

Benefits

Most employees associate employee benefits with health care. Fortunately, employee benefits extend well beyond health benefits and vacations. If your company offers paid or discounted benefits, during your orientation you will learn when and if you qualify for benefits. Benefits may include health care and paid vacations. Depending on the size of the employer, the number of employees, and the hours they work, an employer may or may not be required to provide health care according to the Patient Protection and Affordable Care Act (PPACA), also known as Obamacare. If health care is not offered by your employer, most Americans must obtain health care coverage, secure an exemption, or pay a fee. Typically, only regular, full-time employees are entitled to major benefits. Some large employers allow employees to select which benefits best meet their lifestyle needs. Providing employees their choice of benefits is called a cafeteria plan. Most benefits become effective immediately, whereas others may become effective after employees have successfully completed the introductory period. If you are not certain of when your benefits become effective, consult your human resource department.

If you qualify for benefits, you will be given paperwork to complete. Provide accurate information and keep copies of these forms in a secure place for easy reference. Personal medical information is confidential. Not even your boss should have access to this information. The only people within your company who will know your medical information will be those who are administering your health benefits.

Even if employers are not required to provide healthcare under the PPACA, small employers may offer health benefits as an incentive to attract and retain quality employees. Common health-related benefits include medical, vision, and dental insurance. **Medical benefits** include coverage for physician and hospital visits. Physician coverage sometimes includes chiropractic (bone alignment/massage) and physical therapy (rehabilitation) services. If you or someone in your family utilizes these services, make sure you understand the coverage and stipulations. Research emergency room access and coverage, as well as coverage for pharmaceuticals (prescription drugs). **Vision benefits** include care for your eyes. Some plans pay for eyeglasses only; others pay for contact lenses and/or corrective surgery. Once again, be familiar with your plan and its coverage. **Dental benefits** provide care for your teeth. Check how frequently you are allowed to visit your dentist for routine checkups. Typically, this is twice a year. Identify if your plan pays for cosmetic dental care such as teeth whitening or orthodontia.

Exercise 8.2

Assume you only qualify for four benefits from the following list. Which four would you choose and why?

Chiropractic	Medical	Dental
Vision	Emergency	Company car
Well child care	Day care	Prescriptions
Paid vacation	Life insurance	Retail discounts
Paid holiday	Free parking	Personal days
Paid training	Tuition reimbursement	Mobile communication devices
Family medical	Bonuses	Flexible scheduling

Most people utilize health benefits only when faced with an obvious health issue. Although it is easy to ignore routine checkups, it is in your best interest to practice preventive care. As soon as you become eligible for benefits, schedule an appointment with a physician for a routine physical. Get your vision checked and see your dentist. This is important not only for preventive purposes, but also for establishing relationships with medical professionals. Take advantage of the benefits for which you are eligible, because both you and your employer are paying for them. Finally, this may be a good time to evaluate personal health habits. If you need to shed a few pounds or eliminate a poor habit such as smoking or drinking, now is a good time to change your behavior.

Employers may provide you several choices for health policies. Few policies pay 100 percent of health expenses. Employees typically pay a copayment, or small percentage of the total fee. Many health insurance programs provide a list of medical professionals and health facilities that accept their insurance. Any time you do not use one of these preferred providers, the health insurance will not pay for your care, or it will pay only a percentage of the total. You are responsible for the rest of your bill. When selecting a health program, carefully review the list of providers and facilities for familiarity and convenience. Check emergency access, well child care, preventive care, or any other medical care you may need now or in the future.

Exercise 8.3

What are important considerations for you and your dependents when choosing a physician, hospital, dentist, eye doctor, and pharmacy?

_____

_____

_____

As your benefits are being explained in detail, identify who else in your family may be entitled to these benefits. Frequently, benefits are available to an employee's spouse/partner and/or children. You may have to pay a bit more if you add people to your coverage, but it may be worth the extra cost. As health-care costs continue to increase, you and your family need security and access to quality health care.

Ask if your company offers a retirement plan. A **retirement plan** is a savings plan for when you permanently leave the workforce. If your company offers a retirement plan, join it and start saving. By having funds automatically deducted from your earnings prior to receiving your paycheck, you will most likely not miss the money. Many company-sponsored retirement savings plans are tax deferred, which means you do not pay taxes on these funds until you retire. This provides an added incentive to begin planning for your future.

Payroll, paydays, accrued vacation, and sick days or sick leave are terms used to discuss monetary benefits. Your employee handbook identifies when you are paid. Typically, payday is every two weeks, on the fifteenth and last day of each month. Some companies offer employees the option of receiving payment by a traditional paycheck or by having wages electronically deposited into a bank account. A traditional paycheck is in two parts. One part is the actual check, and the second part of your paycheck is called a pay stub. If you have funds transferred electronically to your bank account, you will receive only a pay stub. This pay stub contains information including hours worked, total pay, taxes paid, and any other deductions that were taken from your paycheck. Any money taken from your original earnings is documented on your pay stub. Keep payroll stubs in a file for tax purposes and future reference. The Internal Revenue Service (IRS) recommends you keep these records on file for three years.

Identify what vacations and holidays your new employer provides and when they will become available to you. Though not required, some employers pay extra wages for working on a holiday. Companies may or may not provide personal days which allow employees to take off without explanation. These days may also be used for personal emergencies. When taking a personal day, provide your employer ample notice prior to taking this time off. Information regarding vacations and holidays should be clearly communicated in the employee handbook or on the company website. If it is not, obtain this information from either your supervisor or the human resource department.

Topic Situation

Joseph's best friend has a family cabin and invited Joseph to spend a long weekend at the lake. Unfortunately, Joseph had been on the job for only two months and had not accumulated any vacation time. Joseph really wanted to go to the cabin. He remembered that the company provides sick leave and one personal day a year. Having been on the job for such a short period of time, Joseph wondered if it would be okay to either call in sick or take a personal day.

TOPIC RESPONSE
What should Joseph do?

If you have a family, are planning a family, or care for an elderly or terminally ill family member, identify company policies regarding pregnancy and family leave. There are laws to protect you from pregnancy discrimination and/or provide relief in these situations. Some employers even provide additional benefits beyond those required by law.

Open-Door Policy

An **open-door policy** communicates to employees that management and the human resource department are available to listen should the employee need to discuss a workplace concern. Think of this policy as more of a "we're here to listen and help" policy. Employers want to know about employee concerns when they first arise so they can be dealt with as soon as possible and prevent the problem from becoming larger or reoccurring. As will be discussed in Chapter 12, the easiest way to deal with conflict is immediately, openly, and honestly. Do not be afraid to speak with your supervisor or the human resource department regarding any workplace issue that causes you concern.

Unions

Depending on the size and nature of your employer, you may have the opportunity to join a union. A **union** is an organization whose purpose is to protect the rights of employees. This organization is a third party that represents you and your colleagues' interests to your employer. The representation is outlined in a contract between the employer and the union, called a **bargaining agreement**. Unions negotiate on behalf of employees on issues related to salaries, benefits, work schedules, performance measures, and other common employee matters. Because the bargaining agreement is a contract that states the rights of employees, it is equally as important as the employee handbook. Employees pay dues (a fee) to a union for union representation. As a bargaining union member, you are trusting that the union will negotiate in your best interest. Employees have the right to unionize (become members of a union), and they also have the right to choose not to join a union. However, depending on the union contract, employees who choose not to be a union member may still have to pay dues if they benefit from the bargaining agreement. Union membership is only for non-management employees.

If your company's employees are represented by a union, the union will contact you and invite you to become a member when you begin employment. Employees who are covered by the union must abide by the bargaining agreement. In addition to the issues addressed earlier, the bargaining agreement also addresses grievance procedures. A grievance is a formal complaint filed by the union when the union contract (bargaining agreement) is not being enforced. Take time to carefully read the contract and keep it in a place where you can easily use it for later reference. Know the names of and how to contact union officials who can assist you with work-related issues.

Your primary union contact will be the shop steward or union representative. This individual is an employee of your company but has agreed to serve as a primary contact between company employees and the union. The shop steward or union representative knows the union contract in great detail and will make every effort to assist you with a work-related issue. Do not hesitate to contact the shop steward or union representative.

Once a new bargaining agreement is presented, union members vote on approving the contract. It is imperative that you exercise your right to vote on a new contract or any union-related issues. The union contract dictates work rules, benefits, and other issues important to your work environment.

Behave responsibly by taking an active role in knowing what services are available to you through the union, the names of union officials who can assist you, and what value and services the union provides through its representation.

MyStudentSuccessLab	Please visit **MyStudentSuccessLab:** Anderson|Bolt, Professionalism Skills for Workplace Success, 4/e for additional activities, resources, and outcomes assessments.

Workplace Dos and Don'ts

Do read and keep the employee handbook for reference	*Don't* ask your boss questions if you have not first referred to the employee handbook
Do utilize the employee handbook to identify paydays and company holidays	*Don't* demonstrate poor planning and be unaware of important days at work
Do maintain documentation for your performance evaluations	*Don't* interrupt during a performance evaluation
Do immediately speak with your boss if there is a conflict regarding your work performance	*Don't* wait until a workplace issue gets out of control to share concerns with your boss

Concept Review and Application

You are a Successful Student if you:

- Explain common HR policies, including employment-at-will and right to revise
- Describe specific considerations when selecting health-related benefits
- Explain the responsibility employees have as union members

Summary of Key Concepts

- The human resource department is responsible for hiring, training, compensation, benefits, performance evaluations, complaints, promotions, and changes in your work status.

- An employee orientation is the time to learn all about a company, its major policies, and employee services.

- The employee handbook is an important document that outlines an employee's agreement with his or her employer regarding work conditions, policies, and benefits.

- Employee benefits may include direct (monetary) benefits and indirect (nonmonetary) benefits such as health care and paid vacations.

- Be aware of and take advantage of benefits that are available to you.

- Be aware of paydays, paid holidays, and sick leave policies.

- Unions are designed to protect the rights of employees.

Self-Quiz MATCHING KEY TERMS

Match the key term to the definition using the identifying number.

Key Terms	Answer	Definitions
Bargaining agreement		1. A formal document provided by the company that outlines an employee's agreement with the employer regarding work conditions, policies, and benefits
Corporate culture		2. An employee with varied work hours and normally does not qualify for employer benefits
Dental benefits		3. A third-party organization that protects the rights of employees and represents employee interests to an employer
Employee handbook		4. A formal appraisal that measures an employee's work performance
Employee orientation		5. Values, expectations, and behavior of people at work; the company's personality reflected through employees' behavior
Employment-at-will		6. An employee that works a pre-determined number of hours per week and is eligible for employer benefits
Full-time employee		7. An employee who is hired for a specified period of time, typically to assist with busy work periods or to temporarily replace an employee on leave
Human resource department		8. Insurance coverage for teeth
Introductory period		9. A statement contained in many employee handbooks that provides an employer the opportunity to change or revise existing policies
Job description		10. Insurance coverage for vision (eye) care

Key Terms	Answer	Definitions
Medical benefits		11. Communicates to employees that management and the human resource department is available to listen should the employee need to discuss a workplace concern
Mentor		12. Insurance coverage for physician and hospital visits
Open-door policy		13. Typically the first one to three months of employment when employers evaluate a new hire's performance and determine if the new hire should continue as an employee (Also known as probation and orientation period)
Part-time employee		14. Policies that do not contractually obligate employees to work for the company for a specified period
Performance evaluation		15. Someone who can help an employee learn more about his or her present position, provide support, and help develop the employee's career
Retirement plan		16. A document that outlines specific job duties and responsibilities for a specific position
Right to revise		17. A savings plan for when an employee permanently leaves the workforce
Temporary employee		18. A department responsible for hiring, training, compensation, benefits, performance evaluations, complaints, promotions, and changes in work status
Union		19. A time when a company provides new employees important information including the company's purpose, structure, major policies, procedures, benefits, and other matters
Vision benefits		20. A contract between an employer and a union that addresses salaries, benefits, working conditions, and other common employee matters

Think Like a Boss

1. How should you handle an employee who keeps coming to you asking for information regarding major policies, vacations, and benefits?

2. How can a boss consistently communicate an open-door policy?

Activities

Activity 8.1

Identify and discuss three typical areas that employers consider in performance evaluations.

1. _____

2. _____

3. _____

Activity 8.2

Research a job description for your target job. Identify common performance criteria for that specific job.

_____

_____

_____

Activity 8.3

How can a potential employee identify if he or she is a "right fit" for a specific company?

_____

_____

_____

Activity 8.4

Assume you are now eligible for health benefits and must choose specific health-care providers. Identify a local physician, eye doctor, dentist, and hospital that you would utilize. Why did you select these providers? How did you go about selecting them?

	Name	Why Selected	How Selected
Physician			
Eye doctor			
Dentist			
Hospital			

9

Communication

trust • honesty • information

After studying these topics, you will benefit by:

- Demonstrating knowledge of the communication process and the impact effective communication has on workplace and career success
- Stating the primary communication media and their appropriate uses
- Considering the importance word choice and effective listening have in verbal communication
- Describing the primary methods of non-verbal communication
- Identifying the appropriate written communication to use in various workplace situations
- Explaining the purpose and process of effective documentation
- Developing the elements of effective presentations

HOW DO YOU RATE?

Have you mastered workplace communication?	Yes	No
1. I do not use foul language.	☐	☐
2. I respect people's personal space.	☐	☐
3. I do not allow emotions to influence my communication.	☐	☐
4. I believe I am a good listener.	☐	☐
5. When appropriate, I send handwritten notes to coworkers.	☐	☐

▶ If you answered "yes" to four or more of these questions, you are well on your way to mastering workplace communication. Communication success begins by presenting your message in a professional manner and focusing on the needs of the receiver.

Communication at Work

Meetings, e-mails, phone calls, texts, presentations, and formal and informal discussions play an important role in business and require proper attention and protocol. Employees who have a basic understanding of how to effectively and appropriately communicate in the workplace are at a significant advantage. Knowing what, when, and how to communicate creates a positive impression on others and helps you achieve your objective. Effective professional and electronic communication is vital to workplace success. This chapter presents the fundamentals of professional communication. Chapter 10 will specifically focus on the appropriate use of electronic communications, including its tools, practices, and protocols.

Workplace Communication and its Channels

Imagine going to work, sitting at your desk, and for one day sending and receiving no communication. If there were no face-to-face contact, no phones, no e-mails, no text messages, no meetings, and no memos to receive or write, business would come to a complete standstill. Even if you are talented at your job, if you cannot communicate with others, you will not keep a job, much less succeed. This chapter discusses the process and importance of effective communication in the workplace and provides information on how to improve workplace communication skills. **Communication** is the process of a sender transmitting a message to a receiver with the purpose of creating mutual understanding.

Improving communication skills is an ongoing process. As explained in Chapter 5, information is power. Therefore, your goal at work is to share appropriate, timely, and accurate information with your boss, your coworkers, and your customers.

| **Topic Situation** |

TOPIC RESPONSE

What type of information do employees need to know?

While eating lunch with employees from other departments, Sarah listened to others complain about how their bosses did such a poor job communicating with them. The employees complained that they never knew what was going on within the company. Sarah had no reason to complain, because she has a manager who makes every effort to share whatever information he knows within the department. After each manager's meeting, Sarah receives an e-mail outlining major topics that were discussed. During Sarah's department meeting, her manager reviews the information a second time and asks his employees if there are any additional questions. Sarah appreciates the fact that the manager enjoys and values communicating important information with his employees.

You have a professional obligation to share timely and relevant information with the appropriate people at work. In the workplace, there are two primary communication channels: formal and informal. **Formal communication** occurs through the formal (official) lines of authority. This includes communication within your immediate department, division, or throughout the company. Formal communication occurs either vertically or horizontally within an organization. Vertical communication either flows down an organizational structure (via written correspondence, policies/procedures, and directives and announcements from management) or up (most commonly through reports, budgets, and requests). Horizontal communication occurs among individuals or departments at the same or close organizational levels.

The second type of communication channel is informal. **Informal communication** occurs among individuals without regard to the formal lines of authority. For example, while eating lunch with friends, you may learn of a new policy. A major element of the informal communication network is called the **grapevine**. The grapevine is an informal network where employees discuss workplace issues of importance. However, it rarely is 100 percent accurate. Although it is important to know about current events at work, do not contribute negative or inaccurate information to the grapevine. Avoid making assumptions if the information is incomplete. When you are aware of the facts, clarify the information. If someone shares information that is harmful to the company or is particularly disturbing to you, you have a responsibility to approach your boss and ask him or her to verify the rumor.

When the grapevine is targeting individuals and their personal lives, it is called **gossip**. Gossip is personal information about individuals that is hurtful and inappropriate. Spreading gossip reflects immaturity and unprofessional behavior and you risk losing credibility with others. Should someone begin sharing gossip with you, politely interrupt and clarify the misinformation when necessary. Respectfully tell the individual that you do not want to hear gossip and/or transition the conversation to a more positive subject. You have a right to defend your coworkers from slander (individuals bad-mouthing others), just as you would expect coworkers to defend you. After a while, your colleagues will learn that you do not tolerate gossip at work and they will reconsider approaching you with gossip.

Refrain from speaking poorly of your coworkers and boss. As a result of human nature, you may not enjoy working with all of your colleagues and bosses. You do not have to like everyone at work, but everyone needs to be treated with respect. Even if someone speaks poorly of you, do not reciprocate the bad behavior. It only displays immaturity on your part and communicates distrust to your colleagues.

The Communication Process

The purpose of communication is to create mutual understanding. Communication is important for maintaining good human relations. Without basic communication skills, processes break down and an organization may collapse.

The communication process (illustrated in Figure 9.1) involves a sender and a receiver. Communication begins with a **sender**, the individual conveying a message. The sender must identify what message needs to be sent and how best to transmit this message. The sender has several options for sending the message. The message can be sent verbally, in written form, or non-verbally. Identifying the specific message and how it will be sent is called **encoding**.

Once the sender encodes the message, the message is sent to a receiver. The **receiver** is an individual that receives and decodes a message. **Decoding** is how the receiver interprets the message. The receiver then sends feedback to

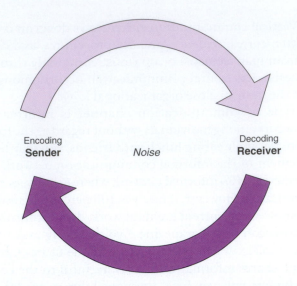

Encoding
Sender   *Noise*   Decoding
                                                        Receiver

Figure 9.1

Communication Process

the sender. **Feedback** is a response to a sender based on the receiver's interpretation of the original message.

Many barriers can hinder the process of creating mutual understanding and successful communication, causing a breakdown in the communication process. The first barrier to overcome is clearly identifying the specific message to be sent. Once the message is identified, the sender needs to determine how best to send (encode) the message in a manner that will be properly interpreted (decoded) by the receiver. If the sender is not a strong communicator, his or her verbal, written, or non-verbal communication may be misinterpreted by the receiver because the message was at risk of being misinterpreted before it was even sent. The receiver contributes to the communication breakdown if he or she incorrectly interprets the message.

Another barrier to effective communication is **noise**. Noise is anything that interrupts or interferes with the communication process. The noise can be audible (you can actually hear it with your ears), or the noise can occur through other senses, such as visual, mental, touch, or smell. Noise may also include emotions such as hurt, anger, joy, sadness, or surprise.

Talk It Out

Identify the noise you experience during class.

Topic Situation

TOPIC RESPONSE

How could Keira have been more open to the speaker?

A supervisor in another department really irritates Keira. Keira has never shared this annoyance with anyone. One day, Keira was asked to attend a meeting led by the irritating supervisor. As Keira sat in the meeting, she had a hard time focusing on the message. Her mind was wandering through mental noise. At the end of the meeting, Keira was embarrassed that there were no notes to share. Dislike for the irritating supervisor affected her ability to listen and be a good receiver. Keira learned a tough lesson that day and made a commitment to be open to every communication, regardless of liking or disliking the sender.

Communication is complete only when all the components of the communication process work together to effectively send the message as they are intended to be sent. In order for this to occur, the sender must select the right medium and overcome noise. The receiver must then be willing to accept the message and provide feedback to acknowledge that the message has been received correctly.

As previously stated, a key element of effective communication is the communication medium (how the message will be sent). Communication media include verbal, non-verbal, and written communication. Let us further explore these three types of communication media.

Verbal Communication and Listening

Verbal communication is the process of using words to send a message. The words you select are extremely important. When you use only basic words in your communications, you risk appearing uneducated or inexperienced. In contrast, when you use a highly developed vocabulary, you may appear intimidating or arrogant. If others do not know the definitions of the words you are using, they will most likely not ask for clarification for fear of appearing ignorant. Therefore, your intended message will fail. When selecting words for your message, identify whether these words are appropriate or if the words can be misinterpreted. Use proper English and grammar. Be as clear as possible in your intent and how you verbally convey your message. When people are nervous or excited, they frequently speak at a rapid pace, increasing the probability that the message will be misinterpreted. Your tone of voice also conveys or creates images. It adds to others' perception of you, which either enforces or detracts from your message.

Successful verbal communication involves **listening**, the act of hearing attentively. Listening occurs not only with our ears, but also through our non-verbal responses. The three primary levels of listening are active listening, passive listening, and not listening at all. **Active listening** is when the receiver provides full attention to the sender without distraction. An active listener will provide frequent positive feedback to the sender through non-verbal gestures such as nodding, eye contact, or other favorable body language. Favorable verbal feedback may also include rephrasing the message to ensure or clarify understanding. With **passive listening**, the receiver is selectively hearing parts of the message and is more focused on responding to what is being said instead of truly listening to the entire message being sent. Passive listening is sometimes called conversational listening. In today's society, we have so many inputs trying to attract our attention that we often get anxious to share our point of view in a conversation and interrupt the sender. Interrupting is rude and disrespectful. Show others respect by not interrupting conversations. If you accidentally interrupt someone, immediately apologize and ask him or her to continue his or her statement. When a receiver fails to make any effort to hear or understand the sender's message, he or she is in **non-listening mode** and is allowing emotions, noise, or preconceptions to impede communication. Sometimes it is obvious the listener is not listening, because he or she either responds inappropriately or does not respond at all.

While the ideal is to consistently be an active listener, we know this is not always possible. However, every effort should be made to strive toward active listening. When you are talking, stop and listen for feedback. Too frequently, a person will have so much to say that he or she does not stop talking long enough to provide the receiver time to respond. The receiver's response is the only way a sender can verify that a message has been properly received.

Web Search

Are you a good communicator? Search the Internet to identify and take an online quiz related to effective communication.

Talk It Out

Name situations where it is easy to be in non-listening mode. What can an individual do to improve his or her listening skills in such situations?

Non-Verbal Communication

Non-verbal communication is what you communicate through body language. Even without uttering a word, you can send a very strong message. Body language includes eye contact, facial expressions, tone of voice, and the positioning of your body. Non-verbal communication also includes the use of silence and space.

An obvious form of body language is eye contact. It is common to look someone in the eye to communicate honesty and sincerity. At other times, looking someone in the eye coupled with a harsh tone of voice and an unfriendly facial expression may imply intimidation. In the United States, those who fail to look someone in the eye risk conveying to their receiver that they are not confident or, worse, are being dishonest. Make eye contact with your audience (individual or group), but do not stare. Staring is considered rude and intimidating. If your direct eye contact is making the receiver uncomfortable, he or she will look away. Be aware of his or her response and adapt your behavior appropriately.

Eye contact is part of the larger communication package of a facial expression. A receiver will find it difficult to interpret your eye contact as sincere and friendly when your message is accompanied by a frown. A smile has immense power and value. On the other hand, make sure you don't smile when listening to someone who is angry or upset. He or she may misinterpret your smile as condescending or as pleased by the distress. As explained previously, when actively listening, a nod implies that you are listening or agreeing with a sender's message. Even the positioning of your head can convey disagreement, confusion, or attentiveness.

Another element of non-verbal communication is the use and positioning of your body. Having your arms crossed in front of your body may be interpreted in several ways. You could be physically cold, angry, or uninterested. When you are not physically cold, having your arms crossed implies that you are creating a barrier between yourself and the other person. To eliminate any miscommunication, it is best to have your arms at your side. Do not hide your hands in your pockets. In speaking with others, be aware of the positioning of your arms and those of your audience. Also, be aware of the positioning of your entire body. Turn your body toward those to whom you are speaking. It is considered rude to turn your back to or ignore someone when he or she is speaking. In this case, you are using your entire body to create a barrier. Avoid this type of rude behavior. This only communicates immaturity on your part.

Exercise 9.1

With a partner, take turns identifying and noting the physical cues for the following emotions: concern, distrust, eagerness, boredom, and self-importance.

_____

_____

_____

The use of your hands is extremely important in effective communication. Through varied positioning, you can use your hands to nonverbally ask someone to stop a behavior, be quiet, or reprimand him or her. Be aware of the positioning of your hands and fingers. In the United States, it is considered rude to point at someone with one finger. Many finger and hand gestures commonly used in the United States are quite offensive in other countries. If you have nervous gestures such as popping your knuckles, biting your nails, or continually tapping your fingers, take steps to eliminate these habits.

Apart from a professional handshake, touching another person at work is not acceptable. People in our society frequently place a hand on another's shoulder as a show of support; however, some interpret that gesture as a threat or sexual advance. Therefore, when at work, keep your hands to yourself.

Proxemics is the study of distance (space) between individuals and is also an important factor in body language. An individual's personal space is about one-and-a-half feet around him or her. The appropriate social space is four feet from an individual. Standing too close may be interpreted as intimidation or may imply intimacy. Neither is appropriate for the workplace. However, distancing yourself too far from someone may imply your unwillingness to communicate. Be aware of the space you allow between you and your receiver. Many variables are involved in effective nonverbal communication. Interpret body language within its entire context. For example, if you are communicating with a colleague with whom you have a positive working relationship and your coworker crosses his or her arms, your coworker is most likely cold. Consider the entire package: environment, relationship, and situation.

Silence is also an effective and powerful communication tool. Silence communicates to your audience that you are listening and are allowing the other party consideration. Not immediately responding to a message provides the sender time to clarify or rephrase a message and provides you time to control your response.

Silence sometimes makes individuals uncomfortable because our society is used to filling up silence with noise. Active listeners take time to digest what is being said and formulate a thoughtful response. An active listener will wait at least three to five seconds before responding. When first using silence, it may feel awkward, but you will quickly discover that you are becoming a better communicator because you are taking time to respond thoughtfully and appropriately. Recognize that there are times when it is appropriate to not speak. In stressful situations, silence is perhaps one of the most important communication tools you possess. Silence can also be a powerful tool when dealing with both conflict and negotiation. As presented in Chapter 5, recognize and respect cultural differences in verbal communication in regard to word use and meaning.

Emotion is another element that affects non-verbal communication. Although reality may cause you to express emotions that are difficult to control, try to control your emotions in public. If you feel you are beginning to cry or have an outburst of anger, excuse yourself. Find a private area and deal with your emotion. When you are crying or distraught, splash water on your face and regain control of your emotions. If you are getting angry, assess why you are angry, control your anger, and then create a strategy to regain control of how best to handle the situation in a professional manner. Any overt display of anger in the workplace is inappropriate, can damage workplace relationships, and could potentially jeopardize your job. When you become emotional at work, you lose your ability to logically deal with situations and risk losing credibility and the trust of others. Practice effective stress management and think before you respond.

Written Communication

Professional writing is a necessary skill for effective workplace communication. **Written communication** is a form of business communication that is printed, handwritten, or sent electronically. Because the receiver of your message will not have verbal and non-verbal assistance in interpreting your written message, take great care to ensure that the correct message is being communicated. You are normally not present when a written message is received; therefore, the receiver will be drawing additional conclusions about you based on the grammar, vocabulary, and presentation used in the written communication.

As you advance in responsibility within an organization, you will be required to conduct an increasing amount of written communication, including formal business letters, memos, and e-mail messages. You may also have the opportunity to communicate through instant messaging, texts, blogs, or wikis, discussed more in Chapter 10. Written business correspondence represents not only your professionalism and intelligence, but also your organizational abilities. Consistently present written correspondence in a professional manner. Ensure all written communication is error-free by proofreading the message prior to sending. Choose words that clearly and concisely communicate your message. The three most common forms of written communication in the workplace are letters, memos, and electronic messages. Written communication in a professional workplace should be typed and not handwritten. An exception to this rule is when you are sending a handwritten note conveying a personal message.

Plan your message for successful written communication. Identify what you want to communicate, to whom you need to communicate, and what desired action you want the reader to take after reading your message. After you have determined what you want to communicate, write a draft that is free of emotion and negativity. Written communication should begin with a professional greeting and end with a complimentary closing. If the purpose of your correspondence is to address a negative situation (e.g., complaint), begin with a positive note and then factually address the situation, but do not attack an individual. Do not send or write any message conveying anger. A good rule of thumb is to always put good news in writing and place negative information in writing only when necessary.

After you have drafted your message and eliminated emotions and negativity, review your correspondence and delete unnecessary words. Keep written correspondence short and simple. Do not be wordy, and minimize personalization words (*I*, *my*) as much as possible. Well-written correspondence not only communicates a core message, but also clearly communicates how the sender wants the reader to respond to the communication. Include contact information and a deadline in your written communication if relevant.

Keep the correspondence simple. Identify and insert words that project a professional image. Know the definitions of the words you are using, and use these words appropriately. A thesaurus is an excellent tool to expand one's vocabulary, but do not overdo it, and be sure to use words in the correct context.

After you have finished writing your message, identify who should receive the message. Share your correspondence only with individuals who need to know the information and, when appropriate, with individuals whom the correspondence affects.

The remainder of this chapter focuses on common written business correspondence, including a business letter, a memo, and a handwritten note. Chapter 10 addresses written communication that occurs through electronic technologies, including e-mail, texting, instant messaging, blogging, and wikis.

Business Letters

A **business letter** is a formal, written form of communication used when your message is being sent to an individual outside of your organization. External audiences may include customers, vendors, suppliers, or members of the community. Although it is still common for formal business letters to be sent through traditional mail, many businesses now send formal business letters as attachments to e-mails. Letters are to be written in proper business format and on company letterhead. Clearly communicate your message and expected follow-up activity to the receiver in a professional and concise manner. Letters sent should be error-free. Proofread, sign, and date the letter before mailing.

Company **letterhead** is quality paper that has the company logo and contact information imprinted on it. Letters sent as attachments in an e-mail should be typed on company letterhead. Figure 9.2 shows the correct business

(Do not type QS and DS, these are shown for correct spacing.)	
Since most business letters will be on letterhead (preprinted business address), you need about a two-inch top margin before entering the current *date*.	August 1, 2018
	QS (4 enters or returns)
The *inside address* should include the title, first, and last name of receiver.	Ms. Suzie Student Word Processing Fun 42 Learn Avenue Fresno, CA 93225
	DS (2 enters or returns)
The *salutation* should have title and last name only.	Dear Ms. Student: *DS*
	The first paragraph of a letter should state the reason for the letter. If you had any previous contact with the receiver, mention it in this paragraph. *DS*
For the *body,* all lines begin at the left margin. Use a colon after the salutation and a comma after the complementary closing.	The second (and possibly a third) paragraph should contain details. All information needing to be communicated should be included here. *DS*
	The last paragraph is used to close the letter. Add information that is needed to clarify anything you said in the letter. Also, add any follow-up or contact information. *DS*
Keep the *closing* simple.	Sincerely, *QS*
	Sarah S. Quirrel
The writer's first and last name should be four enters or returns after the closing to give the *writer* room to sign (remember to have the writer sign).	Sarah S. Quirrel Instructor *DS*
Typist's initials *Enclosure* is used only if you add something in the envelope with the letter.	bt Enclosure

Figure 9.2

Letter Format

August 1, 2018

Ms. Suzie Student
Word Processing Fun
42 Learn Avenue
Fresno, CA 93225

Dear Ms. Student:

It was a pleasure speaking with you over the telephone earlier today. I am delighted that you have agreed to serve as a guest speaker in my Communications class. The purpose of this letter is to confirm the details of the upcoming speaking engagement.

As I mentioned in our conversation, the date for your scheduled lecture is Wednesday, October 14, 2018. The class meets from 6:00 p.m.–8:30 p.m. You may take as much time as you need, but if possible please allow a student question and answer period. There are approximately 60 students, and the classroom contains state-of-the-art technology. If you have specific technology requests, do not hesitate to contact me. Enclosed is a parking permit and map of the campus directing you to the appropriate classroom.

Once again, thank you for continued support of our students. I and my students are looking forward to you sharing your communications insight and expertise with us on October 14. If you have any additional questions, please do not hesitate to contact me via e-mail at S.Quirrel@teaching.com or call me at 123-456-7890.

Sincerely,

Sarah S. Quirrel

Sarah S. Quirrel
Instructor

bt
Enclosure

Figure 9.3

Letter Example

letter format. Figure 9.3 provides an example of a business letter. Please note that a business letter can have various styles; employees should follow the company-preferred style.

When a business letter is not being sent electronically, most companies utilize matching number 10 envelopes (Figure 9.4). Address the envelope with

Exercise 9.2

Practice folding a letter to fit into a number 10 envelope.

_____

_____

_____

S&L Professionalism Corp.
222 Student Success Lane
Kahului, HI 93732

Ms. Suzie Student
Word Processing Fun
42 Learn Avenue
Fresno, CA 93225

Envelope Example

the same information that is in the inside address. Fold the letter in thirds, starting at the bottom and folding up one-third of the way and then fold the top over the bottom, and place it in the envelope with the opening on top.

Business Memos

Business memos (sometimes called interoffice memorandums) are used internally—that is, when the written communication is being sent to a receiver within an organization. Although e-mail is the most common form of internal communication, a traditional business memorandum is still used for internal formal documentation and announcements and is sometimes attached to an e-mail message. A memo includes the receiver's name, sender's name, date, and subject. As with a business letter, include all facts needed to properly communicate the message, but be brief and to the point. Ideally, memos should be no longer than one page. Most word processing software has templates for creating memos.

Figures 9.5 and 9.6 illustrate common business memo formats. As with business letters, many companies have a preferred memo style. Check with your employer to ensure you are utilizing the proper format. A discussion on business e-mail is presented in Chapter 10.

Handwritten Notes

A handwritten note is a personal form of communication. In a professional workplace, it is appropriate to send a handwritten note to acknowledge special events in careers or personal lives (e.g., promotion, birthday, or birth of a child). It is also acceptable to send a handwritten note to encourage a colleague, offer condolences for the loss of a loved one, or to thank someone. Handwritten notes are written in pen on a note card. However, it is also acceptable to acknowledge an occasion with an appropriate greeting card. In some situations it is acceptable to send an electronic thank-you or personal message. Handwritten notes do not need to be lengthy; generally, just a few sentences are sufficient. Acknowledge or encourage coworkers, bosses, and others with whom you work by sending handwritten notes when appropriate.

As mentioned in Chapter 4, a thank-you note is a powerful tool for building relationships. When you express thanks, individuals are more likely to continue assisting and supporting you. Send a thank-you note when someone does

(Do not type DS, these are shown for correct spacing.)	
Start the memo two inches from the top of the page.	**MEMO TO:** Loretta Howerton, Office Manager
	DS
	FROM: Lawrence Schmidt, OA/CIS Trainer
	DS
Double-space after each *heading*. Bold and capitalize only headings, not the information.	**DATE:** January 6, 2018
	DS
Use initial caps in the *subject line*.	**SUBJECT:** Memo Format for Internal Correspondence
	DS
Body—single-space, no tabs, left align. Double-space between paragraphs.	A memorandum is an internal communication that is sent within the organization. It is often the means by which managers correspond with employees and vice versa. Memos provide written records of announcements, requests for action, and policies and procedures. Use first and last names and include the job title.
	DS
	Templates, or preformatted forms, often are used for creating memos. Templates provide a uniform look for company correspondence and save the employee the time of having to design a memo. Word processing software has memo templates that can be customized. Customize the template so it has the company name and your department name at the top. Make sure you change the date format (month, day, year). It should be as it is seen at the beginning of this memo.
	DS
Reference initials (typist's initials) *Attachment notation*, only if needed (if you attach something)	bt Attachment

---

Figure 9.5

Memo Format

MEMO TO: Loretta Howerton, Office Manager

FROM: Lawrence Schmidt, OA/CIS Trainer

DATE: January 6, 2018

SUBJECT: Accounting Department Computer Training

This memo is to confirm that the computer training for the accounting department will occur on February 1, 2018, in the large conference room. Although the training is scheduled from 9:00 a.m.–11:30 a.m., I have reserved the room for the entire morning, beginning at 7:00 a.m.

As we discussed last week, this may be a good opportunity to offer breakfast to the department prior to the training. If this is something you would like to pursue, please let me know by next Tuesday, and I will make the proper arrangements. Thank you again for the opportunity to provide computer training to your team.

bt

Figure 9.6

Memo Example

Include the date.	*June 3, 2018*
Start your note with a salutation and the receiver's name.	*Dear Ms. McCombs,*
Be brief but specific about why you are thanking the person. Include how you benefited from the person's kindness. Do not begin every sentence with *I*.	*Thank you for loaning me your book on business etiquette. I especially liked the chapter on social events and dining. Your constant encouragement and mentoring mean so much to me.*

Sincerely, |
| Use a complimentary closing, and do not forget to sign your name. | *Mason Yang* |

Figure 9.7

Thank-You Note

something for you that takes more than five minutes or when someone gives you a gift. Deliver the note as soon as possible. Handwritten thank-you notes are commonly sent to an interviewer after an interview. This will be further discussed in Chapter 15. Figure 9.7 displays the correct format and key elements of a handwritten note.

Documentation

Documentation is a formal record of events or activities. Some industries require documentation to track a project's progress or an employee's time for client billing. Documentation may be necessary for an employee evaluation, for advancement, in an instance in which a policy is not enforced, or when an abnormal event has occurred that has the potential to evolve into conflict at a later date. These events may support performance issues, business relationships, and business operations. It is not necessary to record every event that occurs at work. Employees should maintain a file of positive feedback received from coworkers, supervisors, or customers. Employees should also document relevant negative business situations, such as a workplace injury, an angry customer, or an employee conflict, in order to protect themselves and/or the employer. Although there are numerous methods of documenting and retaining important information and events, the basic elements to be recorded remain the same.

Depending on the purpose of your documentation, effective documentation records the who, what, when, where, and why of a situation. Effective documentation essentials include who was present when the event occurred and how witnesses to the event behaved or responded. When describing what happened, keep the documentation factual and do not allow feelings and assumptions to distort the facts. Include the date, time, and location of the event. Documentation can be kept electronically, in a journal, or through minimal notations on a calendar. If the documentation is for billing or client purposes, your employer will provide the documentation format. Whatever system you choose, keep your documentation in a secure, private location along with copies of supporting memos, letters, or other communications. If you are ever called on to defend your actions, you will have the ability to easily gather pertinent information.

Presentations

Formal and informal presentations are normal workplace events, and sometime in your career you may be asked to give a presentation. Being prepared and professional will help increase audience interest and reception. Presentations are rich in media and may include written, verbal, visual, and/or nonverbal communication. A successful presentation begins with a goal. Identify the purpose of your presentation and ensure that every word, visual aid, activity, and/or handout will support the overall goal of the presentation. After the purpose of the presentation has been identified, an outline of key points should be identified to reinforce the message you want individuals to respond to or remember.

Formal presentations include three elements: the verbal content, the visual content, and support content. Verbal content provides the primary message, visual content summarizes the message, and support content reinforces the message. Verbal content includes the detailed information you wish to share with the audience. When presenting, do not read directly from the visual content. Speak clearly and at a normal pace using professional and appropriate language. Face your audience. If you are using a screen, keep your back toward the screen. Do not block the audience's view of your visual. Beware of both verbal and non-verbal gestures. Nothing will distract an audience quicker than an overuse of "um," "like," and "you know." Hands in pockets, crossed arms, or tapping feet are examples of distracting physical gestures. Dress professionally, and do not wear anything that may distract from your message.

Visual content includes anything the audience will view or any activity the audience will perform during your presentation. Often this involves some type of technology, including presentation software, videos, or music. When using presentation software, do not overdo the use of graphics, color, or animations. Test all equipment and software prior to the actual presentation to ensure the equipment is working and the software is compatible. Preparation and practice ensure that your visual content and/or activities are the appropriate length. If you are including your audience in an activity (e.g., game), make directions simple and the activity brief. Keep your audience focused, and do not allow the activity to distract from your message.

Support content normally comes in the form of a handout. This is an excellent way to reinforce your verbal and visual message in writing. A popular format for a handout allows the audience to fill in the blanks as you present your message. Add non-distracting professional and visual appeal to your handout. As you create your handout, follow the same order as the presentation outline. Check your visual presentation and support materials for clarity, spelling, and grammatical errors. When you are certain your support content is error-free and professional, make enough copies for each member of your audience.

Formal presentations are an excellent way to increase workplace credibility and individual confidence. Successful presentations are a result of planning. Remember, practice makes perfect.

Slang and Foul Language

Slang is an informal language used among a particular group. Although different generations, cultures, and technology use slang, avoid using slang in the workplace to ensure your message is not misinterpreted by others. Become a more effective communicator in the workplace by eliminating the use of slang.

Your words reflect what is going on in your heart and mind. There is no appropriate time to use profane and offensive language at work. Even in times of stress or at social functions, you are representing your company and must do so in a professional manner. Practice self-control. Attempt to eliminate foul or offensive language from your personal and professional vocabulary. If you utilize inappropriate language at work, immediately apologize. Make a mental note of what situation caused you to behave poorly and learn from the experience. Ask yourself how you could have better handled the situation, and mentally rehearse a proper, more acceptable method of verbally handling a challenging situation.

Think About It

What slang terms do you use in text messages that may be offensive to others?

Potentially Offensive Names

Names that could be considered sexist and offensive are inappropriate in a business setting. These include names such as honey, sweetie, and sexy. Using inappropriate names toward coworkers could expose you and your company to a potential sexual harassment lawsuit. Even if the individual being called these names acts as if he or she is not offended, the person may actually be offended or insulted but is afraid to tell you. Eliminate these words from your workplace vocabulary. In addition, do not use gender-specific titles when referring to certain jobs. For example:

Instead of	Use
Postman	Postal Carrier
Policeman	Police Officer
Waitress	Server
Stewardess	Flight Attendant
Maid	Housekeeper

Not Always About You

There is one word that often dominates written and verbal communication but frequently turns receivers off. Unfortunately, too often, the sender is unaware of its overuse. The word is *I*. Be cautious with the use of this word. Self-centered people use it to draw attention; whereas those who lack self-confidence may subconsciously use the word to protect themselves. Many individuals who overuse this word may not know how to turn the conversation to others, so they choose to stay in a safety zone. When you are using verbal communication, think before you speak. When writing, if your initial sentence includes *I*, try to rephrase your message. Prior to sending written correspondence, review your message and reduce the number of sentences that begin with the word *I*.

Exercise 9.3

Take five minutes and interview a classmate about college and his or her career choice. While you are getting to know each other, keep track of how many times your new friend says the word *I*.

_____

_____

_____

MyStudentSuccessLab Please visit **MyStudentSuccessLab**: Anderson|Bolt, Professionalism Skills for Workplace Success, 4/e for additional activities, resources, and outcomes assessments.

Workplace Dos and Don'ts

Do carefully think through your message and the appropriate medium	*Don't* be in such a hurry to send your message that an incorrect message is sent
Do demonstrate professionalism in the formatting, word choice, and grammar in your written communication	*Don't* write and send messages when you are angry
Do express kindness to others with both your words and body language	*Don't* utilize foul language at work or at home

Concept Review and Application

You are a Successful Student if you:

- Demonstrate proper formatting for a business letter and memo
- Write a handwritten thank-you note
- Create a documentation record
- Explain the difference between gossip and the grapevine

Summary of Key Concepts

- Effective communication is necessary for workplace success.

- The goal of communication is to create a mutual understanding between the sender and the receiver.

- There are appropriate times to utilize both the formal and informal communication channels.

- The communication process involves a sender, a receiver, noise, and feedback.

- Listening and silence are effective tools for effective communication.

- Thoughtfully consider the right words to increase the chance of successful written and verbal communication.

- Because the receiver of your message will not have verbal and non-verbal assistance in interpreting your message, take great care with all written messages.

Self-Quiz MATCHING KEY TERMS

Match the key term to the definition using the identifying number.

Key Terms	Answer	Definitions
Active listening		1. Communication that occurs through formal lines of authority
Business letter		2. The process of using words to send a message
Business memos		3. When a receiver is selectively hearing parts of a message and is more focused on responding
Communication		4. A formal record of events or activities
Decoding		5. The process of a sender transmitting a message to a receiver with the purpose of creating mutual understanding
Documentation		6. An informal network where employees discuss workplace issues of importance
Encoding		7. Quality paper that has the company logo and contact information printed on it
Feedback		8. Communicating through body language
Formal communication		9. How the receiver interprets a message
Gossip		10. When the grapevine is targeting individuals and their personal lives
Grapevine		11. The study of distance (space) between individuals

Key Terms	Answer	Definitions
Informal communication		12. Written communication sent within an organization
Letterhead		13. Communication that occurs among individuals without regard to formal lines of authority
Listening		14. A form of business communication that is printed, handwritten, or sent electronically
Noise		15. An informal language used among a particular group
Non-listening mode		16. The act of hearing attentively
Non-verbal communication		17. A response to a sender's message
Passive listening		18. The individual conveying a message
Proxemics		19. A formal, written form of communication sent to individuals outside of an organization
Receiver		20. When a receiver fails to make any effort to hear or understand the sender's message
Sender		21. Anything that interrupts or interferes with the communication process
Slang		22. An individual that receives and decodes a message
Verbal communication		23. When the receiver provides full attention to the sender without distraction
Written communication		24. Identifying a specific message and how it will be sent

Think Like a Boss

1. One of your employees uses bad grammar that is reflecting poorly on your department's performance. How can you get a handle on this problem?

2. Employees keep saying they do not know what is going on at work. What steps would you take to increase workplace communication?

Activities

Activity 9.1

Without infringing on someone's privacy, discreetly observe a stranger's body language for approximately five minutes. Stay far enough away to not hear him or her speak. Name at least two assumptions you can make by simply watching the person's gestures, movements, and expressions.

Gesture, Movement, or Expression	Assumption

Activity 9.2

Watch a television news show for a half hour. Document at least two facial expressions of an individual being interviewed. Did the individual's facial expressions match his or her statements?

Facial Expression	Statements

Activity 9.3

Review the following letter. Identify five formatting errors and correct those errors.

April

Sandra Wong, Vice President
Human Resource Department
Robinson Enterprises
55123 W. Robinson Lane
Prosperity, CA 99923

Dear Sandra Wong

It was a pleasure speaking with you this afternoon regarding the average salary you pay your receptionists. This data will be useful as our company begins creating a new receptionist position for our California site.

I am most appreciative of your offer to mail me a copy of your most recent salary guide for all production positions. I look forward to receiving that guide in the mail. As a thank you for your kindness, I am enclosing coupons for our company product.

If there is any information I can provide to assist you, please let me know. Thank you again for your cooperation.

Sincerely,

Cory Kringle

List Errors	Correction
1.	
2.	
3.	
4.	
5.	

Activity 9.4

Review the following memo. Identify five errors (not including spacing) and make the appropriate corrections.

MEMORANDUM

Re: Budget Meeting

To: Mason Jared

From: Cory Kringle

Date: May 1

Hey Mason. I wanted to remind you that we have a meeting next week to talk about next year's budget. Bring some numbers and we'll work through them. Bye.

-Cory

List Errors	Correction
1.	
2.	
3.	
4.	
5.	

Activity 9.5

You received a letter of recommendation from a teacher. Write a thank-you note.

_____

_____

_____

After studying these topics, you will benefit by:

- Explaining the professional use of electronic communication tools
- Constructing and utilizing e-mail messages
- Utilizing phone etiquette
- Demonstrating the proper use of portable devices and texting
- Applying professionalism to social media tools
- Identifying proper behaviors in video and teleconferences

HOW DO YOU RATE?

Are you addicted to an electronic device?	Yes	No
1. Within five minutes of waking, do you check your device for messages?	☐	☐
2. Do you have to check your device at least once every hour?	☐	☐
3. Do you use your device in locations or situations where you know it is not appropriate?	☐	☐
4. Do you always have your device visible or easily accessible?	☐	☐
5. Are you unable to go an entire day without access to your device?	☐	☐

▶ If you answered "yes" to two or more of these questions, you may be addicted to your electronic device.

Electronic Communications at Work

We live in a multitasking, fast-paced world that has resulted in technology addiction. The traditional workplace of the past has evolved into a virtual workplace where most people are electronically connected. Today's workplace communicates through venues including phones, e-mail, mobile devices, texting, instant messaging, blogs, wikis, and audio and video conferencing. The more we are connected technologically, the greater the opportunity for disconnected messages. As a complement to Chapter 9, this chapter focuses on electronic communications in the workplace. Due to the frequency and speed of message transmission, those who communicate through today's virtual workplace need to take great care to ensure all electronic communications are sent in a clear, concise, and professional manner.

Telecommunication Basics

With the growth of technology in the workplace, the proper use of electronic communication tools, devices, and equipment becomes increasingly important. Common communication tools include various forms of computers, software, e-mail, Internet, and mobile (smart) devices. Employers may provide these tools to employees free of charge. If you utilize company-provided tools (including a computer, company server, or e-mail address), the tools, equipment, and messages are to be used strictly for company business. This includes the use of the Internet and electronic messaging. Many organizations have a **technology use policy** that outlines expectations including privacy, liability, and potential misconduct issues related to the use of company technology. When utilizing company technology, favorably represent the company by not violating confidentiality and by using the equipment for work purposes only. There is growing concern regarding the security of company-owned technology devices and the confidentiality of the information these devices contain. Take special care with the security of company equipment, where you store it, and how it is used.

With a wide variety of electronic device options, keep in mind that there are proper times and places for their use. In some work situations, it is perfectly appropriate to use a laptop or mobile device. In other situations, it is highly inappropriate. Only utilize the communication tool when it is relevant to the discussion or issue you are addressing. The communication tool should not distract from the conversation at hand. When in doubt, ask permission to use the device and explain why you want to use it to assist in the discussion.

It is not productive to constantly be checking these venues for messages; however, it is also unproductive to ignore messages. The primary communication venue varies by job; a general rule of thumb is to check your workplace communication devices at least twice a day (more frequently if it is part of your job duties). Messages should be responded to when they are read.

Practice good computer hygiene. Just as you would not show up to work when you are sick, you do not want to be responsible for contaminating others' communication tools when sharing information electronically. Although businesses often have company-wide computer security systems, it is a good practice to routinely scan your equipment for viruses, cookies, and other malicious

Talk It Out

Identify the impact technology will have on the workplace in the next five years.

TOPIC RESPONSE

How might Kathy have better handled the situation of using her smart phone during the meeting?

Kathy was in a company meeting. During the meeting, there was disagreement on whether the company's competitor had specific information on its website. Kathy quickly pulled out her smartphone and began pulling up the competitor's website. One of the company executives glared at Kathy, assuming she was being rude by texting or tending to personal business. Catching the executive's glare, Kathy immediately held up the device and said, "I don't want to appear rude. I am quickly checking our competitor's site." Kathy quickly retrieved and reported on the site and was able to contribute valuable information to the discussion.

coding that can be potentially harmful. Review the e-mail address prior to opening a message from an unknown recipient. If the address or subject line is suspicious, contact your computer administrator to ensure the message is not a virus. Regularly back up documents for preservation should a storage device fail.

Business E-Mails

Electronic mail (e-mail) is the most common form of internal and external electronic communication in the workplace. With e-mail, you can directly type a message or attach a business document. E-mail creates more efficient communication within an organization and with individuals outside of the organization.

When sending an e-mail, ensure the subject line clearly and concisely describes the purpose of the message. Include a brief descriptive statement in the subject line that makes the receiver want to read the message. Do not leave the subject line blank or use the words "Hi" or "Hello"; these may be mistaken for spam or a virus. Avoid using the words "Urgent," "Important," or "Test" in a subject line. If the message is urgent or important, use the priority tag (commonly an exclamation point). Marking all messages as urgent weakens your credibility, as it becomes hard for individuals to identify which of your messages truly are urgent. People may stop reading your messages immediately or altogether. A proper business e-mail subject line is formatted the same as a hard-copy memo subject line, which uses initial capitalization of words and no abbreviations.

As with all workplace equipment, business e-mail should be used only for business purposes. Emoticons (faces made and embedded in e-mail messages) and decorative backgrounds are inappropriate in business messages. Refrain from forwarding messages that are not work-related. These non-business messages clutter company servers and may contain viruses and cookies. Maintain an organized and updated electronic address book and make every attempt to preserve the confidentiality of your address book.

When you receive a work-related message that requires a reply, respond to the message. Ignoring a message is rude and communicates to the sender that you do not care. Ignoring messages also puts you at risk of being excluded from future messages.

Writing E-Mail Messages

E-mail is a necessary technology in nearly every workplace and can be easily misused. As with formal correspondence written on company letterhead, an e-mail should utilize proper layout, spelling, and grammar. Just like writing a business letter, composing a successful e-mail message involves planning and identifying the purpose of your message. The message may be informational, it may be a topic for discussion, or it may require a decision. Clearly state your specific message and what action you want the receiver(s) to take. Doing so increases the likelihood of successful communication.

Identify who should receive your e-mail message and include only individuals who need to know the information you are sharing. It is not necessary to include your boss in every e-mail. Any individuals to whom the message is directed should be listed in the "To:" line. You can also courtesy copy (Cc:) the message to other individuals by listing them in the "Cc:" line. Individuals who are named in or affected by the message that are not included in the "To:" line are to be listed in the "Cc:" line. E-mail software has the option of blind copying (Bcc:) your message to others by including their addresses in the "Bcc:" line. When an individual is blind copied on an e-mail, the "Bcc:" recipients can see the main and "Cc:" recipients, but the main and "Cc:" recipients do not see the "Bcc:" recipient(s). Therefore, not all recipients are aware of who is included in the message, which can create a sense of mistrust. The use of blind copying is discouraged, except in the case of sending an e-mail to a mailing list where you do not want the recipients to see the other names due to privacy issues.

Exercise 10.1

Your boss (Penny@workspace.star) asks you to send a copy of a meeting memo to your coworkers: Jennifer@workspace.star, Julie@workspace.star, and Gene@workspace.star. Complete the entries.

To:	
Cc:	
Bcc:	
Subject:	
Message:	

After you have planned your message, write a draft, clearly communicating your primary message early in the e-mail so as to capture and keep the reader's attention. Include the key points you want to communicate and the specific action you are requesting from the reader(s). Consider the reader's perspective and communicate the message in a positive manner. If your message contains several points, bullet or number each item and/or use subheadings to make it easier for the reader to follow and properly respond to your message. After you have finished composing your draft, edit the message. Delete unnecessary words and review the message for clarity and conciseness. People often judge others' professionalism based upon their writing skills. When you are satisfied, proofread the entire message. Most business e-mail software contains both spelling and grammar check. Use these tools. Nothing lessens credibility faster than receiving a message filled with spelling and grammatical errors. If your message refers to an attachment, do not forget to include the attachment. Review your message for clarity, ensure the proper file is attached (if relevant), and confirm that you are sending the message to the appropriate parties. Finally, make certain that the subject line concisely summarizes your message. After you have taken these steps, send your e-mail.

When sending e-mail, practice the following positive habits:

- E-mail messages written in all capital letters or with large and colorful letters are interpreted as yelling and are considered rude.
- If your e-mail software has the ability to embed a permanent signature, use it. Include your first and last name, title, company, business address, contact phone, and e-mail address.
- Common e-mail software has the capability of requesting a return receipt whenever a message is received and read. Some individuals consider this an invasion of privacy. Use this function only when necessary.
- If you will be out of the office and unable to access and/or respond to your e-mail within a reasonable time, utilize an automated response to all e-mail messages informing the senders that you are unavailable. Remember to retract the automated response when you return.

Forwarding e-mail messages is a common practice in the workplace, as it saves time and brings parties into the loop on a subject they may have not originally been involved with. If misused, however, this practice can cause conflict and/or embarrassment. When forwarding messages, include only individuals for whom the information is relevant. Prior to forwarding a message, ensure that none of the earlier information in the string of previous e-mail messages could embarrass anyone or contains confidential information. If you are unsure if the information has the potential to embarrass someone, do not forward the message. Simply summarize the situation in a new e-mail with new recipients and copy (Cc:) the original parties, if appropriate.

Talk It Out

When is it appropriate to use the return receipt feature in an e-mail message?

Topic Situation

TOPIC RESPONSE

Did Patrick appropriately handle the coworker's negative e-mail? Why or why not?

Using business e-mail was a common activity for Patrick. Patrick was careful to always include an appropriate subject line, ensured that the content was professionally and concisely written, requested an action or follow-up activity, and sent it to the appropriate individuals. Patrick was taken aback one day when a coworker sent Patrick a negative e-mail for including inappropriate recipients in an e-mail message. The individual scolding Patrick had sent his negative e-mail to everyone in the department, which embarrassed Patrick. Patrick reviewed the e-mail in question and did not see anything wrong with the original message or the recipient list. As Patrick reflected on how best to respond, Patrick decided that the individual who sent the negative message acted on emotion and embarrassed himself to all of his coworkers in the process of trying to embarrass Patrick. Therefore, Patrick felt it best to not respond.

Portable Devices and Texting

Today's work environment relies on current technologies to improve communication. This is achieved through the use of portable communication devices. Common devices include smartphones, portable music/entertainment devices, tablets, laptops, and netbooks. Though the use of these business tools is acceptable in most business situations, employees need to display proper etiquette when using these devices. Just as it is impolite to verbally interrupt someone who is talking, it is also impolite to interrupt a conversation or meeting with incoming or outgoing electronic communications.

There are two basic guidelines for using personal electronic communication devices at work. First, you may use your communication device if you are alone, in a private area, and its use is permitted. Second, use your device when attending a meeting or business activity and its use is necessary for communication. If the use of the device is not relevant to the activity, silence or turn off your device and put it away. If you forget to turn off the sound and the device rings, do not answer it. Apologize and immediately silence the device or turn it off. In the rare situation that you are expecting an important call that you must answer, if possible, inform the leader of the meeting, briefly explain the situation, and apologize ahead of time for the potential interruption. Should you receive a message, politely excuse yourself from the room before answering and take the message in private. Although these guidelines are for business purposes, they should pertain to personal use as well. Please review the information regarding telecommunication etiquette detailed in Chapter 4.

In some situations, texting is a valuable communication tool. In the workplace, texting should be used only for brief, informal communications. When you are in the presence of others, a general rule of thumb is to text only if the message is short and related to the business at hand. For example, if you are negotiating a deal, you may text your boss to identify terms to present. Prior to texting, inform those present of your activity. Just as with all written business communication, when texting for business purposes, the use of proper spelling and grammar is essential. Many people utilize text slang, text shorthand, acronyms, and codes in personal e-mails and texts. The use of these styles is not appropriate for business communications.

Unnecessary and constant texting has become a habit for many. Just as with other portable communication devices, it is not appropriate and is considered rude behavior to view and send text messages while with others (including discreetly during meetings). Therefore, when in meetings, turn off or silence and put your device out of sight unless it is explicitly necessary for the meeting. If not, the mere presence of the device may be tempting and divert your attention from the business at hand. It is also rude to use your communication device while dining or attending public meetings or performances. It is also not polite to take calls in front of others. This implies that the individuals you are with are not important. When taking a call, apologize for the interruption, excuse yourself, and step away for privacy.

It is also inappropriate to use or display portable music/entertainment devices in the workplace unless the device provides quiet background music appropriate for a professional workplace and it does not disturb others. If you work alone and wish to use personal headphones, first check with your boss to ensure doing so is allowed and does not pose a safety hazard.

Talk It Out

What should you say to someone who is inappropriately using his or her mobile device during a meeting?

Phone Etiquette

The phone is one of the most common workplace communication tools. Phone etiquette, whether landline or wireless, is something every individual must practice to create and maintain a professional image for his or her company. Because the individual(s) on the other end of the phone cannot see you, proper communication through word choice, tone, pitch, and rate of speech is of great importance.

When answering a call, try to answer on the first or second ring. Start with a greeting such as "Good morning," and identify yourself and the company. Convey a positive, professional attitude when speaking on the phone. Smile when you speak to create a friendly tone. Speak clearly and slowly, and do not speak too softly or too loudly. If you take a second call and need to place the first caller on hold, politely tell the individual on the phone that you are placing him or her on hold. If an individual is placed on hold for more than one minute, get back on the line and ask if you can return the call at a later time.

Taking a call without explanation in the presence of others implies that the individual in your presence is not important. When engaged in a conversation with others, allow the call to go into voice mail. If you are expecting an important call, before starting a conversation, state that you are expecting an important call and will need to take it when it arrives. When the call is received, politely excuse yourself. If you are in your office, politely ask your office guest to excuse you while you take the call.

When making a phone call, identify yourself to the receiver and ensure the receiver has time to talk. If you expect the discussion to be lengthy and you did not prearrange the call, ask the receiver if he or she has time to talk or if there is a more convenient time. When you are part of a phone conversation, do not eat or tend to personal matters. Speaker phones are useful communication tools for specific situations and also require proper etiquette. A speaker phone should be used only when you are part of a conference call with other participants in the same room or when you require a hands-free device. Use a speaker phone only when you are in a private room where your call will not be distracting to others in your work area. When you use a speaker phone, ask individuals included in the call for permission to use the speaker phone. Alert those included in the call that others are in the room with you and make introductions. This ensures confidentiality and open communication between all parties. Those using a speaker phone should be aware that any small noise they make may be heard and distracting to those on the other end of the line.

Voice mail messages are a part of business communication. A voice mail impression is equally as important as communicating in person. When leaving a voice mail message, keep the message brief and professional. State your name and the purpose of the call, and leave a return number at the beginning of the message. Speak slowly and clearly and leave a short but concise message. After you have left your message, repeat your name and return number a second time before ending the call. Promptly return voice mail messages and routinely check and empty your voice mail box.

On both portable and landline phones, keep your voice mail greeting professional. Include your name and the company name in the message. Clever voice mail greetings are not professional. Musical introductions or bad jokes do not form favorable impressions when employers or customers are attempting to contact you.

Exercise 10.2

You are the account clerk at Garret and Danielle Accounting. Create a professional voice mail greeting for your company-supplied smartphone.

_____

_____

_____

Social Media Tools

Companies commonly use social media tools such as LinkedIn, Facebook, Instagram, blogs, microblogs, and others for marketing purposes. Some companies hire professionals to maintain and manage their image through social media outlets. Though it may be tempting to post a video or vent about an irate customer, coworker, or administrator online, such behavior is not only unprofessional, but could be a violation of the company's technology use policy. The behavior could also pose potential legal issues for both you and your employer. An increasing number of employers consider any employee use of social media that reflects poorly on the employer as a violation of its technology use policy. Individuals using social media for personal reasons need to separate personal sharing from professional sharing. Many organizations regard the posting of company-related information by employees as divulging confidential or competitive information. Regardless of your company's policy, it is best to refrain from identifying and/or speaking poorly of the company, employees, vendors, and customers in all social media communications.

In addition to e-mail, a growing number of companies are utilizing wikis, blogs, and instant messaging for both internal and external communications. A wiki is a collaborative website where users have the ability to edit and contribute to the site. Blogs, also called web logs, are online journals where readers are often allowed to comment. Instant messaging (IM) is a form of online communication that occurs between two or more parties in real time. Business etiquette regarding the use of these communication methods is similar to that of e-mail. When at work, use these venues only for business purposes. Proper spelling and grammar and clear and concise communications are necessary. As with all forms of written communication, professionalism and tone matter. View your participation in a wiki as a form of teamwork. When making edits to the wiki, be sensitive to how others are receiving your comments and, in turn, accept the suggestions of others. Your goal is to provide an accurate webpage that properly communicates your message. Business blogs are used as both marketing and education tools. The purpose of a blog is to create and enhance relationships, so keep blog posts and comments positive and meaningful. The difference between IM and e-mail is that you are able to identify who is online at the same time you are. Utilize IM only for brief business interactions. Though it is tempting to IM individuals when you see they are online, remember that IM at work is not intended as a workplace social tool. You do not want to become disruptive or annoying when utilizing IM. Whatever electronic communication venue you utilize, remember that you are representing your company.

Web Search

Are you addicted to social media? Conduct a web search to identify and take an online quiz to determine if you are addicted to social media.

It is perfectly common and acceptable to utilize social media tools for personal reasons; however, remember to maintain a positive and professional image, or online identity. An **online identity** is the image formed when someone is communicating and/or researching you through electronic means. Routinely conduct an Internet search of yourself to ensure you have a clean online identity. If there are negative photos, videos, blogs, or other information that reflect poorly on you, have them removed. Maintain a professional electronic personality by only utilizing electronic communications in a professional manner.

Topic Situation

TOPIC RESPONSE

If you were Irene and you knew your new boss saw the social media site open on your computer, how would you respond?

Fran's friend Irene recently got a new job in sales that required her to train and job-shadow her manager for the first few weeks. Fran knew Irene was addicted to both texting and her social media site, so Fran was glad that she now had something to keep her mind focused. During a sales call, Irene was not focused and was using a company laptop to play on her social media site instead of reviewing sales figures. Irene and her client stepped away from the conference table for a minute, and Irene's boss tried to quickly retrieve a figure from the computer. Unfortunately, the boss only saw Irene's social media site. To make matters worse, the site contained a photo of Irene in a crazy pose outside of her new company headquarters, which included the company's name in the picture.

Video and Teleconferencing

In some situations, meetings take place using telecommunications. This form of meeting is referred to as a **conference call** and is treated like a face-to-face, prearranged meeting. A **teleconference** is an interactive communication that connects individuals through a phone line without the opportunity of seeing all participants. It is also common for meetings to take place through video venues such as Skype, WebEx, and Google Talk. A **video conference** is an interactive communication using two-way video and audio technology. It allows individuals in another location to see and hear all meeting participants. A computer, a webcam, and a reliable Internet connection and sometimes a phone line are needed when participating in a video conference. A quiet location is also necessary. When taking part in a video or teleconference, the participant will receive a designated time and specific instructions on how to establish connection. In addition to following the phone interview guidelines presented in Chapter 15, the meeting participant needs to prepare for and treat the telecommunication meeting as if it were a face-to-face meeting. Follow these basic tips for a successful electronic meeting:

- Plan ahead. Research the venue you will be using to address any unforeseen issue. If possible, arrange a premeeting trial to ensure all equipment works properly (including your connections, volume, and microphone).
- Dress professionally if you are visible to other participants. As with face-to-face meetings, visual impressions matter.
- Maintain a professional environment. Conduct your meeting in a quiet and appropriate location. When you are visible to other participants, a bedroom, public place, or outside location is not appropriate.

- Speak to the camera (if you are participating in a video conference). Focus on the webcam as if you were speaking directly to the other participants. Without interrupting or distracting others, feel free to ask questions, take notes, and use hand gestures.
- Avoid distracting noises. Turn off music or any other items that create distracting noises.
- Do not eat or drink during the meeting.
- When teleconferencing, state your name each time you speak. For example, prior to contributing, say, "Hi, this is Ted. I would like to provide a status report on the Phoenix project." Because virtual meetings require a special emphasis on listening, be quiet when others are speaking and do not do anything distracting. Take your turn speaking and do not interrupt. As with face-to-face meetings, which will be discussed in Chapter 11, be prepared and actively contribute.

The number of technology-related workplace tools continues to grow, as do their applications. Though our means of communicating at work may change, the need for professional communication remains the same. Be respectful and concise in your communication and represent both yourself and your organization in a professional manner.

MyStudentSuccessLab Please visit **MyStudentSuccessLab**: Anderson|Bolt, Professionalism Skills for Workplace Success, 4/e for additional activities, resources, and outcomes assessments.

Workplace Dos and Don'ts

Do utilize company technology tools only for company business	*Don't* violate your company's technology use policy
Do practice good computer hygiene	*Don't* forget to routinely scan your computer for viruses and other malicious software
Do recognize the appropriate time and place for workplace technologies	*Don't* allow technology to distract from business matters
Do demonstrate professionalism in business e-mail and texts	*Don't* become addicted to workplace technologies by sharing inappropriate messages
Do display professional meeting habits in video conferencing and teleconferencing	*Don't* let not being face-to-face in a video or teleconference interfere with practicing professionalism

Concept Review and Application

You are a Successful Student if you:

- Write a professional email
- Summarize professionalism regarding electronic communications
- Write a professional text

Summary of Key Concepts

- Send electronic communications in a clear and professional manner.
- Many organizations have technology use policies that address privacy, liability, and potential misconduct issues.
- Do not forward messages at work that do not involve work-related issues.
- When texting for business purposes, the use of proper spelling and grammar is essential
- When utilizing social media for personal use, refrain from identifying your company and/or speaking poorly of the company and/or its customers.
- Maintain a clean online identity.
- Practice good meeting habits in technology-based meetings.

Self-Quiz MATCHING KEY TERMS

Match the key term to the definition using the identifying number.

Key Terms	Answer	Definitions
Conference call		1. An interactive communication using two-way video and audio technology
Online identity		2. Interactive communication that connects individuals through a phone line without the opportunity of seeing all participants
Technology use policy		3. Meetings utilizing technology
Teleconference		4. Outlines expectations including privacy, liability, and potential misconduct issues related to the use of company technology
Video conference		5. Image formed when someone is communicating and/or researching you through electronic means

Think Like a Boss

1. One of your employees has been sending personal texts during meetings. How should you handle this issue?
2. Many employees are taking photos and/or videos at department meetings and company events. Should you be concerned? Why? As the boss, what should you do?

Activities

Activity 10.1

Research current technologies that are being used in your target industry. How can this technology be abused?

New Technology	Potential Abuse

Activity 10.2

Check your online identity by conducting an Internet search on yourself. Is there anything you need to change?

_____

_____

Activity 10.3

Research the most current mobile devices to identify features and how necessary these features are for your target career.

_____

_____

_____

_____

Activity 10.4

Research technology use policies and report on the use of technology, privacy issues, and ownership issues.

_____

_____

Activity 10.5

You are out of the office with a client who is unhappy. Write a text to your boss.

_____

_____

11

Motivation, Leadership, and Teams

unity • energy • accomplishment

After studying these topics, you will benefit by:

- Defining motivation and explaining common motivational factors
- Explaining the primary leadership styles and key qualities of a successful leader
- Examining the difference between leadership and management
- Describing a team, the elements of effective teams, and how they affect performance
- Identifying characteristics of effective team players
- Demonstrating how to deal with difficult team members
- Listing and describing the elements of a successful meeting

HOW DO YOU RATE?

What kind of team member are you?	Yes	No
1. Coworkers would say that I consistently behave in a professional manner at work.	☐	☐
2. I normally arrive at least five minutes early to meetings.	☐	☐
3. When participating in team projects, I always complete my portion of the project on time.	☐	☐
4. If there is conflict within the team, I work to resolve the team conflict.	☐	☐
5. I consistently behave as a leader in work-related situations, including knowing when to lead and when to follow.	☐	☐

▶ If you answered "yes" to two or more of these questions, congratulations. You display both leadership and positive team member behaviors.

A Foundation for Performance

The three elements of motivation, leadership, and teamwork create the foundation for productivity and organizational success. This chapter focuses on these important workplace concepts by first addressing the topic of motivation and explaining how an understanding of basic motivation theories assists you in becoming a more productive employee and leader. A leadership title is not necessary to exhibit leadership behaviors. All employees should strive to display the characteristics of a leader. Motivation and leadership are elements of teamwork. Understanding these three intertwined topics will improve your success.

Motivation

Motivation is an internal drive that causes people to behave a certain way to meet a need. Behavior changes until desired needs are met. You are most likely taking this course because you believe the information you receive will assist you in achieving your career goal. The need to achieve your career goal motivates you to succeed in this course. Motivation is essential for employees to perform their jobs in a quality and productive manner. Several factors can contribute to employee motivation, with the most obvious factor being money. However, monetary payment often may not be the primary motivating factor. In Chapter 1, you created a life plan that outlined personal and professional goals. The timelines you attached to each goal help you monitor your success. Every time you accomplish a goal, you are motivated to achieve more.

Because motivation is an internal drive, people are motivated by different needs. **Maslow's Hierarchy of Needs** (see Figure 11.1) states that throughout one's lifetime, an individual's needs are met as they progress up a pyramid (hierarchy) of five needs. These needs are physiological, safety, social, self-esteem, and self-actualization. Organizational behaviorists have adapted Maslow's hierarchy to a typical workplace. Maslow's lowest level, physiological needs, are physical survival needs. In the workplace, this translates to basic wages. People work to receive a paycheck, which is used for food, clothing, and shelter. The next level in the pyramid is safety needs. In the workplace, individuals desire a safe working environment, job security, and benefits. Only after individuals receive basic wages and job security do they invest in positive workplace relationships, thus reaching the social needs level. After experiencing acceptance by others, individuals move to the self-esteem level of the pyramid. An employee's self-esteem needs are met when they feel they are accepted and valued by others, such as recognition through promotions, degrees, and awards. The final stage of the hierarchy is self-actualization. When employees have achieved meaningful work, most will want to assist others in meeting their needs. They do so by becoming mentors or by finding other means of helping others achieve their career goals.

Maslow's theory is still used today to help managers understand what motivates people. Each level of the hierarchy provides insights and opportunities to motivate individuals. Recognize that not everyone is motivated by the same factors, nor do others have the same needs as you. Observe other employees' behaviors and words, then try to determine what need requires fulfillment. Once you identify others' needs, you can assist in creating an environment that helps meet these particular needs.

Talk It Out

What motivates you to perform at work or at school?

Figure 11.1

Maslow's Hierarchy
of Needs

Maslow in the Workplace

Self-Actualization—Expand Skills

Esteem—Recognition/Respect

Social—Informal Groups

Safety—Job Security/Environment

Physiological—Basic Wages

Other popular motivational theories are David McClelland's Theory of Needs and Victor Vroom's Expectancy Theory. **McClelland's Theory of Needs** holds that people are primarily motivated by one of three factors: achievement, power, and affiliation. Achievement is based on doing something better, the need for power is based on one's desire to influence others, and the need for affiliation is based on an individual's need to maintain positive relationships with others. Although some are motivated by all three needs, most people have a tendency to favor one need over the other two. **Victor Vroom's Expectancy Theory** holds that individuals will behave in a certain manner based on an expected outcome. For example, you may be motivated to study because you expect to perform well on an exam if you study. The expected outcome is a good exam score as a result of studying.

There are many additional motivational theories, but the three presented here provide a basic understanding of the concept of motivation. As you assume either a formal or informal leadership position, your understanding of motivation will assist you in helping both you and others achieve peak performance.

Exercise 11.1

Evaluate the following comments and determine what need on Maslow's hierarchy is being expressed. Refer to Figure 11.1.

Comment	Need Expressed
I have done a similar project in the past; can I help you?	
I need a raise this year.	
Anyone want to join me for lunch?	
I received a sales award. Would you like to see it?	

Motivation is an internal drive, meaning you motivate yourself. Others can only provide a motivating environment. If you find yourself having an unproductive day, take time out and identify what situation placed you in a demotivating frame of mind. There are some workplace situations that you cannot control. In these instances, try to identify what element of the situation you can control and take action. Review your goals to refocus on the big picture, place yourself in a positive frame of mind, and get back to being productive. You are the only one who can control your attitude and your desire to perform well.

Leadership

When reference is made to leaders, many people think of managers. However, there is a difference between a leader and a manager. A manager's job is to plan, organize, and control; a leader's job is to inspire and guide. Although some individuals are formally managers, every employee has the potential to be a leader and should display leadership. **Leadership** is a process of a person guiding one or more individuals toward a specific goal. A leader does not require a title to lead. Leaders motivate others through relationships. At work, these relationships are based on vision, trust, and mutual respect. A leader is one who guides and motivates others. In other words, a leader will help others want to accomplish a job successfully. There are three primary leadership styles: autocratic, democratic, and laissez-faire leadership. **Autocratic leaders** are authoritarian, meaning they make decisions on their own and tell others what to do. **Democratic leaders** make decisions based on input from others. People are encouraged to share ideas. The democratic leader will evaluate the ideas and come to a decision that reflects the collective ideas.

Laissez-faire leaders allow team members to make their own decisions without input from the leader. Employees have complete freedom in making decisions concerning their work. At first glance, most would think a laissez-faire leader is using the best leadership style, but there is no best leadership style. The appropriate leadership style is dependent on the situation, employee skill level, and several other variables. There are some situations that require one individual to tell others exactly what to do, and there are other situations where individuals work best without specific instruction. Historically, leaders used a "command and control" leadership style where they told others what to do without guiding and motivating. Today's leaders collaborate whenever possible to increase employee participation, accountability, and creativity.

An individual cannot lead without followers, and people will not follow a leader they do not trust or respect. Effective leaders display characteristics that make them stand out from others by being positive and supportive of others. These characteristics include providing others with a vision, solving issues in a creative manner, and displaying consistent ethical behavior. Leaders are also excellent communicators and are committed to continual learning. Although these characteristics are not developed overnight, they are worthy of pursuing not only at work, but in other areas of life. Commit to assuming a leadership role whenever possible. Know your work and its purpose, and be someone others can trust by showing others mutual respect. Problem solve, form and communicate a plan, then follow up on projects. Perform your job to the best of your ability. A leader encourages others to succeed. If you are a leader, groom others and provide them the opportunity to lead.

Talk It Out

If the room were on fire, how would each type of leader direct his or her employees?

Web Search

Conduct a web search to identify a quiz that identifies your leadership style.

Leaders understand the importance of accountability and maintaining positive workplace relationships. Effective leaders do not do all the work themselves. When appropriate, they delegate. **Delegating** is when a manager or leader assigns part or all of a project to someone else. As a leader, empower, teach, and mentor others. Ultimately, leaders take responsibility. Become a model by taking your responsibilities seriously and consistently performing in a quality manner.

Becoming a Leader

At work you may be assigned a leadership position by your boss, by a team, or simply because no one else wants to lead. No matter how a leadership position is obtained, be willing and prepared to lead. Begin preparing today by learning new skills, joining committees, training, and attending workshops. Volunteer to serve on a team and learn what skills are necessary to be a successful team leader. Observe successful leaders and/or find a mentor to help you develop strong leadership skills. Learning new skills, including improved communications, will enable you to better think and behave like a leader.

An important element of professional success includes getting involved in your community. One excellent method of increasing leadership skills away from work is to share your expertise with a local non-profit organization. Find a creative way to share your time and talent with those less fortunate or with those who are just learning a skill you have already mastered. Many non-profit organizations are in need of individuals to serve on their board of directors, head projects, or perform specific activities. Volunteer experiences provide excellent opportunities to develop both leadership and team-building skills. Assisting others stimulates creativity, relieves stress, and is also an excellent way to meet and network with others. Additional benefits from volunteerism include an expanded professional network and potential job leads, as will be discussed in Chapter 13.

Exercise 11.2

Name three non-profit organizations that can benefit from your expertise, time, or talent.

_____

_____

_____

One final issue influencing today's leaders is dealing with gender, generational, and cultural differences. Today's leaders represent the contemporary workplace of various ages, religions, gender differences, and other factors that were discussed in Chapter 5. A good leader will remember that not everyone shares his or her beliefs and values. The job of the leader is to unite individuals and use diversity as an asset to successfully achieve company goals. As a leader, be aware of differences among individuals and ensure that your personal actions are in no way offensive or demeaning to others (intentionally or unintentionally). In addition, display zero tolerance for prejudice and avoid stereotypes by treating everyone equally.

Teams and Performance

Most individuals have experienced being part of a successful team. Perhaps it was a sports team, or maybe it was in school when a group of students successfully completed a big project. Whatever the task, individuals were part of a unit whose members all shared and contributed to a common goal and respected one another. These important factors resulted in success. Learning to successfully work alongside others is a skill that is necessary in the workplace. You will most likely be working with others as part of a team. A **team** is a group of people linked to a common purpose. Each team should strive toward creating synergy. **Synergy** is defined as two or more individuals working together and producing more than the sum of their individual efforts. When people are truly working together as a team, performance is at a premium and the result exceeds what each individual can accomplish alone. This section discusses teamwork, factors vital for team success, and the impact teams have on an organization's overall performance.

It is becoming increasingly common for companies to rely on teams to accomplish goals. When workplace teams are assigned the task of reaching a goal, members are accountable to one another and to the organization as a whole. Teams provide opportunities for members to take a leadership role in helping them become accountable for successfully reaching a goal. In a team setting, each member has a sense of ownership for the team's performance. This can occur only when team members are active participants and are accountable to fellow team members.

Several types of teams exist at work. Teams that occur within the organizational structure are formal teams. **Formal teams** are developed within the formal organizational structure and include functional teams (e.g., individuals from the same department) or cross-functional teams (individuals from different departments). **Informal teams** are composed of individuals who get together outside of the formal organizational structure to accomplish a goal. Examples of informal teams include a company softball team and a group of coworkers collecting food for a local charity. Another type of team that is common in today's workplace is the **virtual team**, which functions through electronic communications because the members are geographically dispersed. It is quite common for virtual teams to operate in various time zones and across national borders. Effective communication and premeeting planning is essential when working in a virtual team. Refer to Chapters 9 and 10 for additional information on effective communication and appropriate use of communication technologies.

No matter what type of team situation you are involved with, positive relationships among all team members is necessary for success. Getting the job done depends on a team effort. A team composed of individuals who behave professionally performs better.

Teams go through five stages of team development: forming, storming, norming, performing, and adjourning. In the **forming stage**, individuals are getting to know and form initial opinions about team members. Assumptions are based on first impressions. Sometimes these impressions are right; other times, they are wrong. In stage two, the **storming stage**, some members begin to have conflict with each other. When team members accept other members for who they are (i.e., overcome the conflict), the team has moved into the **norming stage**. It is only then that the team is able to enter the **performing stage**, where they begin working on the task. Once the team has completed its task, it is in the **adjourning stage**, which brings closure to the project. Note that it is normal for a team to experience each of these phases. As a team member, expect and accept when your team is moving through each phase. Some teams successfully and rapidly move through the forming, storming, and norming phases and get right to work (performing), whereas other teams cannot get beyond the initial phases of forming or storming. Make every effort to move your team along to the performing stage, and recognize that minor conflicts are a part of team development. Successful teams will move beyond conflict and accept each member for his or her unique talents and skills.

Although it is common to work with team members you know, there will be situations when you may be part of a team of people you have never met before. Some team members may be from your immediate department (functional teams), some may be from outside of your department (cross-functional teams), and some may be from outside of the company. Strong people skills and a willingness to lead are what make individuals valuable team members.

Characteristics of an Effective Team Member

Common team projects include improving product quality, providing excellent customer service, and creating and/or maintaining company records. An effective team member is one who does his or her job in a manner that is contributing to the project's goal. This requires effective team members to be trustworthy, efficient, and communicate at all times.

As a team member, know the goal and objectives. The activity performed for your team should support the team's goal and objectives. In Chapter 1, you learned how to create personal goals. In Chapter 7, you learned that organizations also create and utilize goals and objectives as part of its strategic plan. When teams work on a specific project, there is a goal. Therefore, the first step for effective teams is to identify a goal and objectives to reach that goal. Do not just jump into a project without a clear understanding of the expected outcome. Do not reinvent the wheel or waste time and money. The best way to avoid these common mistakes is to first review the history of how the situation (problem or opportunity) evolved and then solicit ideas and input from all team members on how to solve the problem or seize the opportunity.

Once the team has identified its goal and objectives, the team can determine how best to successfully achieve the goal. One popular way of doing this is through brainstorming. **Brainstorming** is a problem-solving method that involves identifying alternatives by allowing members to freely add ideas while other members withhold comments. Brainstorming is successful because it is fast and provides members the opportunity to contribute different and creative ideas. Brainstorming starts with the presentation of a problem, such as how to improve office communication. Members then have a set time to make any suggestion for improving office communication. The suggestions can be obvious (e.g., a newsletter) or fun and creative (have a daily off-site office party). No matter the suggestion, members are to withhold comment and judgment on the idea until the brainstorming session is over. In effective brainstorming sessions, even an off-the-wall comment such as the suggestion to have a daily off-site office party may spark a more practical idea that contributes to solving the problem (e.g., have a company-wide gathering or a daily two-minute meeting for communication and motivation).

An effective team member is able to work with everyone on the team. There may be a time you will have to work with someone you don't like. Although conflict is a natural part of team development, occasionally there are teams that are filled with hard-to-resolve conflict. In Chapter 12, you will learn various ways to overcome conflict. Do not allow one member to ruin the synergy of a team. If possible, confidentially pull the member aside and ask that person what it is about the team or project that he or she finds objectionable and how he or she feels the issue can best be resolved. Calmly and logically help the individual work through the issues. Accept the fact that he or she may not (1) recognize there is a problem, (2) openly share the reason for the conflict, or (3) want to come to a solution. If you or others on the team attempt to solve the problem with the difficult team member and he or she rejects the effort, the team needs to move forward without that person. One team member should not have so much power that he or she negatively affects the efforts of the entire team. Although team conflict is a natural stage of team development, it should not cripple a team. When the problem team member is not around, do not allow other team members to talk negatively about the individual. Your job is to be a productive, positive team member who assists in successfully accomplishing the team's goal.

Exercise 11.3

Brainstorm as many ideas as possible to help you and/or your classmates save money while attending college.

_____

_____

_____

After you know the goal of the project and your specific roles and responsibilities in the team effort, know the responsibilities of all the other team members. Whenever possible, identify ways to support other team members and assist them in accomplishing the team's objectives. Take responsibility to attend all team meetings and be on time. Participation, sharing, support, understanding, and concern are all part of serving on a team. During team meetings, be involved in discussions and determine what work is necessary for accomplishing the goal. Do not be afraid to speak during team meetings. Some of your ideas may not be considered, but that does not mean they are not important. Be responsible by finishing assignments in a timely manner. As a team, review all aspects of the project together before adjourning to ensure the project goal and all objectives were met.

Topic Situation

Mason's department was having problems meeting its production goal. The manager asked selected members of the department to form a team and create a plan for increasing production. Mason volunteered to serve on this team. It was the first team project Mason had been involved in, and he did not know what to expect. Fortunately, Mason had a good team leader. The leader sat down with the team and helped identify a goal and objectives so that members knew exactly what needed to be done and who would be responsible for each activity. At the next meeting, the team leader led a brainstorming session. Some good ideas were shared. Mason had an idea but was afraid to speak up, thinking the idea might not be as good as the others.

TOPIC RESPONSE

Should Mason share his idea? How should he respond if his idea is rejected?

Communication is a key element of effective teamwork. Do not make assumptions about others or a team project. If you have questions regarding any aspect of a project, respectfully speak up. If others do not agree with your ideas, keep a positive attitude. If your team made a wrong decision, do not waste time on blame; take corrective action and learn from any mistakes that are made. Each team member should be able to freely state his or her position and ideas. The entire team should then decide which ideas to use. Do not assume that any team member's idea is not worth hearing. The whole point of a team project is to solicit as many ideas as possible to identify the best solution for reaching the goal. If the team makes a decision with which you did not agree, you have expressed and explained your objection, and the team does not agree with your objection, support the team's decision and keep assisting the team in achieving its goal. Conflict is a normal part of teamwork. Learn to work through conflict by practicing open, honest, and timely communication with all team members.

In a team situation, you will usually have your own job to perform, but you are also accountable to fellow team members. The success of others depends on how you do your job. Although you may be working independently from your team members, it is still important that you complete your job on time and correctly. Be an active participant and do not allow team members to do your work because you know they will do it for you. Sometimes in a team situation, one member contributes nothing because that member knows others will do his or her work. If you have a lazy team member, continue to do your best and try to work around that person. Try talking to the poorly performing team member and identify why he or she is not doing his or her share. If the team member provides a good reason, suggest that he or she excuse himself or herself from the team. If the team member simply refuses to perform, you may need to decide as a team to have the unproductive team member dismissed or replaced.

Exercise 11.4

Name the two most important characteristics you would want to see in your team members and explain how these characteristics contribute to team success.

_____

_____

_____

Meetings

A common form of team interaction and workplace communication is a meeting. Meetings are either informational, discussion driven, decisional, or some combination of the three. Meetings can be formal or informal. The most common form of meetings in the workplace is a department meeting, which is when a boss formally meets with his or her employees.

Prior to meetings, a meeting agenda is normally distributed to all attendees. A **meeting agenda** is an outline of major topics and activities that are scheduled to be addressed during the meeting. Some agendas have time limits attached to each item. If you receive an agenda prior to a meeting, take time to read the agenda and become familiar with the topics of discussion. If there is an item you would like placed on the agenda, notify the person in charge of the meeting. If you are responsible for an agenda item, plan what you are going to share and/or request prior to the meeting. Prepare handouts for each attendee if necessary.

The most common type of meeting is a face-to-face meeting where all parties are physically present in one location. Arrive early to face-to-face meetings. Depending on the size of the meeting, there can be one table, or many tables filled with meeting attendees. If there is a head table, do not sit at the head table unless you are invited to do so. If there are no assigned seats and you are speaking, sit toward the front of the room. The **meeting chair** is the individual who is in charge of the meeting and has prepared the agenda. This person normally sits at the head of the table. If the chair has an assistant, the assistant will usually sit at the right side of the meeting chair. Other individuals in authority may sit toward the front of the table or at the opposite end of the table. If you are unsure of where to sit at a conference table, wait to see where others sit and then fill in an empty seat.

It is not only important, but respectful for individuals to show up on time and be prepared for meetings. Most formal business meetings will follow some form of **Robert's Rules of Order**, a guide to running meetings. Robert's Rules of Order is often referred to as parliamentary procedure. At the start of a meeting, the meeting chair will call a meeting to order and, if appropriate, review the minutes from the last meeting. After the review of minutes, the meeting chair will ask that the minutes be approved, then the agenda issues will be addressed in order. At the close of the meeting, the meeting chair will adjourn the meeting.

As a meeting participant, take your turn speaking by contributing thoughtful and relevant information when appropriate. Keep your discussion to the agenda topic and assist the meeting chair by keeping the discussion moving, with all contributions being professional, respectful, and focused on the goals of the company.

When distance separates meeting participants, virtual meetings take place. These meetings occur through the use of technology such as a telephone, e-mail, or videoconferencing. A more in-depth discussion on professional behavior related to these electronic communication venues is presented in Chapter 10.

Team Presentations

Some work situations require employees to create and provide a presentation as a team. The presentation material presented in Chapter 9 also applies to team presentations. When working with others, the first step is for all team members to agree on the presentation goal. The team should then create a presentation outline. Using the outline as a foundation, discuss and agree on the verbal, visual, and support content. Just as with other team situations, each member needs to take responsibility and be accountable to one another. Do not just split up sections and piece a presentation together at the last minute. Team presentations must be completed and reviewed by the entire team before presenting. Demonstrating positive human relations skills is a key to the success of a team. Each member must communicate, share duties, and behave in a respectful and professional manner.

> **Talk It Out**
>
> What do students dislike most about team presentations?

MyStudentSuccessLab | Please visit **MyStudentSuccessLab**: Anderson|Bolt, Professionalism Skills for Workplace Success, 4/e for additional activities, resources, and outcomes assessments.

Workplace Dos and Don'ts

Do be an active participant by being accountable to fellow team members	*Don't* ignore the needs of others in the workplace
Do be a good team member by being trustworthy and efficient, communicating at all times	*Don't* leave the leadership process up to others
Do express your thoughts during team meetings	*Don't* think your ideas are of no value
Do recognize that people are motivated by different factors	*Don't* ignore team meetings and deadlines
Do make every effort to increase your leadership skills	*Don't* allow negative team members to disrupt the team's performance

Concept Review and Application

You are a Successful Student if you:

- Summarize how motivation influences you, your coworkers, and your company's success
- Create and implement a plan to enhance your leadership skills
- Explain how to create a productive team

Summary of Key Concepts

- Motivation is an internal drive that causes you to behave a certain way to meet a need.
- Everyone has the ability to become a successful leader.

- Most companies use teams to accomplish goals.
- An effective team comprises individuals who share a goal and respect for one another.
- A good team member is one who does his or her job in a manner that is productive toward the end project.
- Although team conflict is a natural stage of team development, do not allow conflict to cripple a team.
- Communication is a key element of effective teamwork.

Self-Quiz MATCHING KEY TERMS

Match the key term to the definition using the identifying number.

Key Terms	Answer	Definitions
Adjourning stage		1. Authoritarian leaders who make decisions on their own and tell others what to do
Autocratic leaders		2. A theory that individuals will behave in a certain manner based on an expected outcome
Brainstorming		3. When the team focuses on achieving the task
Delegating		4. Leaders who allow members to make their own decisions without input from the leader
Democratic leaders		5. Teams that function through electronic communications because they are geographically dispersed
Formal teams		6. The individual in charge of a meeting
Forming stage		7. An internal drive that causes people to behave a certain way to meet a need
Informal teams		8. A group of people linked to a common purpose
Laissez-faire leaders		9. A problem-solving method that involves identifying alternatives freely
Leadership		10. When the team has completed its task and brings closure to the project
Maslow's Hierarchy of Needs		11. Stage of team development when members get to know each other
McClelland's Theory of Needs		12. Process of a person guiding one or more individuals toward a specific goal
Meeting agenda		13. When a manager or leader assigns part or all of a project to someone else
Meeting chair		14. Leaders who make decisions based on input from others
Motivation		15. Individuals who get together outside of the formal structure to accomplish a goal
Norming stage		16. A guide to running meetings

Key Terms	Answer	Definitions
Performing stage		17. A theory stating that throughout one's lifetime, individuals progress up a pyramid of needs
Robert's Rules of Order		18. An outline of major topics to be addressed during a meeting
Storming stage		19. Two or more individuals working together and producing more than the sum of their individual efforts
Synergy		20. Teams that are developed within the formal organizational structure
Team		21. When team members experience conflict
Virtual teams		22. When team members accept other members and overcome conflict
Victor Vroom's Expectancy Theory		23. A theory that people are motivated by three factors: achievement, power, and affiliation

Think Like a Boss

1. You have assembled a group of employees to reach the goal of improving customer service in your department, but all they do is argue when they meet. What should you do?

2. Your employees have successfully met their production goals this week. Based on Maslow's Hierarchy of Needs, how can you motivate them to meet next week's goals?

Activities

Activity 11.1

If you were teaching this class, what specific topics or activities would you include in the course to motivate students in each level of Maslow's Hierarchy?

Level	Motivation Factor
Self-Actualization	
Self-Esteem	
Social	
Safety	
Physiological	

Activity 11.2

Research President Abraham Lincoln and answer the following questions (include your source/citations). What key leadership qualities made him unique?

_____

_____

What challenges did he face?

_____

_____

How can you apply lessons learned from your research to your leadership development?

_____

_____

Activity 11.3

Identify common prejudice/stereotypes you have observed that are displayed against the following groups. How should a leader handle the situation?

	Prejudice/Stereotype	**How Leader Should Handle Situation**
Age		
Gender		
Culture		
Disability		

Activity 11.4

As a team member, what specific action(s) can you take to move your team to the next stage?

Forming	
Storming	
Norming	
Performing	
Adjourning	

Activity 11.5

What type of team member are you when it comes to team presentations?

_____

What improvements can you make to become a valued team member?

_____

Activity 11.6

Research two rules from Robert's Rules of Order and share your findings.

Rule	What You Learned
1.	
2.	

12

Conflict and Negotiation

perspective • agreement • rights

After studying these topics, you will benefit by:

- Understanding conflict and how best to respond to conflict at work
- Explaining the various conflict management styles and their appropriate application
- Defining negotiation and applying negotiation techniques
- Identifying harassment and workplace bullying and demonstrating how to respond
- Stating employee rights in the workplace
- Demonstrating how to resolve conflict in both a union and non-union environment
- Recognizing warning signs and proactive steps to take against workplace violence

HOW DO YOU RATE?

How do you handle conflict, negotiation, and harassment?	True	False
1. If I have a conflict with my boss, it is best to immediately go to the human resource department and file a grievance.	☐	☐
2. A successful negotiation results in me always getting my way.	☐	☐
3. A successful strategy for combating workplace bullies is to publicly bully them back.	☐	☐
4. Names such as "honey," "babe," and "darling" are acceptable at work if the coworker knows I'm kidding.	☐	☐
5. If I disagree with someone, I should keep the disagreement to myself and not cause conflict.	☐	☐

▶ If you answered "true" to two or more of these questions, you have a tremendous opportunity to learn more about conflict management. Employees have a right to stand up for themselves without violating the rights of others. As you'll learn in this chapter, a conflict should be resolved as soon as possible, at the lowest level possible.

Conflict

No workplace is without conflict. The key to successful conflict management is knowing how to appropriately handle the conflict in a manner that reflects well on both you and your organization. Although most individuals regard conflict as a negative experience, it can result in a positive outcome if you approach it with the right attitude. This chapter addresses the issue of conflict and its impact on performance. Various methods of dealing with conflict, in addition to tips on how to deal with difficult people, are also presented. Finally, the issues of harassment, workplace violence, and negotiation are discussed.

Conflict is a disagreement or tension between two or more parties (individuals or groups). This disagreement or tension is the result of a perceived threat to one's needs, interests, or concerns. The perceived threat is often based on an individual jumping to conclusions or making assumptions. We generally do not experience major conflict with those of whom we have positive relationships because we are not threatened by these individuals. When we are working with individuals we either do not know or do not have favorable relationships with, we may feel threatened and become defensive when they have different perspectives than ours. Although conflict at work cannot be avoided, you can control your reaction to the conflict.

Resolving Conflict

Conflict is a result of individuals looking at a situation from different perspectives, which reflect diversity of thought. Therefore, do not make conflict personal. If you view conflict as a breakdown in communication, work on overcoming the problem instead of finding fault or blame. Frame the conflict around an issue or situation, not a person. How an individual deals with conflict reflects his or her attitude, maturity level, and self-confidence.

When you find yourself coming into conflict with someone, take a moment and ask yourself if you are making assumptions about the individual and/or situation. Avoid making assumptions by clarifying both facts and your understanding of the situation. Be honest with yourself and recognize your feelings. If you feel the party with whom you are in conflict with is a threat to you, identify why. You may find the source of the conflict is your own insecurity and a result of false assumptions. Conflict resolution begins with you, your attitude toward the other party, and your ability to identify whether or not a conflict truly exists.

Topic Situation

TOPIC RESPONSE
How did Luis mishandle the situation? What assumptions did Luis make about Anthony? How should Luis correct the situation?

Luis's boss asked him to head a team. This was Luis's first experience leading a team and he was excited. As Luis identified individuals that needed to serve on the team, he was unsure about including his coworker, Anthony. He needed Anthony's expertise; however, Anthony has been rude toward Luis in the past. Luis decided to do the right thing and included Anthony in an e-mail he sent all prospective members. Within minutes of sending the message, Anthony responded to everyone on the e-mail list with several questions. Luis immediately responded in a professional manner, but sat at his desk a bit angry about Anthony's message. After a few minutes, Luis reread Anthony's e-mail and realized that the questions were actually good questions addressing issues Luis should have included in the original message.

When someone disagrees with you or hurts you, a natural tendency is to become angry. A common reaction to anger is to retaliate or get even. Unfortunately, this behavior does not reflect that of an individual who is striving to become a logical, mature professional. Follow these basic rules when dealing with conflict:

- Whenever possible, resolve the conflict in person.
- Remain calm and unemotional.
- Be silent and listen.
- Try to view the disagreement from the other person's perspective.
- Explain your position and offer a solution.
- Come to a solution.

Today's technologically connected workplace is resulting in increased reliance on brief, written communications. The digital communication medium opens the door to misinterpretation of messages, which can create conflict. As soon as you recognize a potential electronic miscommunication with another party, try to resolve the conflict face-to-face. Doing so allows you the opportunity to more quickly clarify understanding through both verbal and non-verbal communication. As presented in Chapter 10, it is unprofessional to create and support a negative string of written messages. These types of messages leave a permanent record of negative behavior for all involved parties.

Rarely, if ever, does anyone win when people respond to conflict in anger. An individual who becomes emotional has difficulty resolving the issue in a logical manner. When confronted with conflict, remain calm and unemotional. Acknowledge your hurt feelings or anger, but do not allow emotions to dominate your response. It is easier to view the disagreement from another perspective when your mind is clear. Before responding, attempt to identify why the other person behaved the way he or she did. Slow down and look for the facts and feelings in the message you received. Doing so will help you identify if you misinterpreted the message because of an emotional response or a miscommunication of a fact.

After thinking about the disagreement from the other person's perspective, calmly and rationally explain your position along with a proposed solution. This is the step that could easily lead to an argument if you become emotional. While you are explaining your position, the other party may interrupt and state his or her position. Do not argue. Allow the other party to speak while you remain quiet and listen. This is tough, because we want to defend our opinions. Respond in a mature, professional manner. When the other party is finished speaking, take your turn. If he or she again interrupts, politely ask if you can take your turn responding. In your conversation, look for common ground. Identify the true source of the disagreement and remember to not make the disagreement personal. Try to provide several alternatives to solving the problem, and then agree on a solution. In the workplace there may be situations where both parties need to agree to disagree because a solution cannot be reached.

The following list offers several basic concepts for dealing with conflict in the workplace:

- Only you can control how you respond to a situation.
- Do not allow feelings to dictate actions. Remain calm and unemotional.
- Attempt to resolve a conflict immediately; work with the other party.
- Accept responsibility for your actions and apologize if necessary.
- Retaliation (getting even) is not the answer.
- Do not make the conflict a public issue. Keep the conflict confidential.

If the conflict is negatively affecting your job performance:

- Document the offensive or inappropriate behavior regarding the conflict.
- Seek assistance within the company to resolve the conflict. Whenever possible, begin with the supervisor.
- If an internal remedy cannot be reached, seek outside assistance.

Exercise 12.1

Engage in a conversation with a classmate about the best way to take notes in class. Each classmate should attempt to interrupt the other. Practice handling the interruption in an adult, mature manner. What should you say or do when interrupted?

_____

_____

_____

Conflict Management and Negotiation

Several conflict management styles exist, including forcing, avoiding, accommodating, compromising, or collaborating. The appropriate style to use depends on the situation.

If a behavior is offensive or unacceptable, use the **forcing conflict management style**. This style attempts to make the other party do things your way. The other party has no say whatsoever. This style deals directly with the issue. When using this style, remain calm and unemotional and do not turn the discussion into an argument. Your goal is to provide your solution to the problem while communicating that the inappropriate behavior is unacceptable. The **avoiding conflict management style** is used when you do not want to deal with the conflict, so you ignore the offense. Sometimes we avoid a conflict because the offense is not a big enough deal to upset others. Other times, we avoid the conflict because we are not strong or confident enough to stand up for our rights. Although it is perfectly appropriate to ignore a conflict at times, do stand up for your rights when necessary.

The **accommodating conflict management style** allows the other party to have his or her way without the other party knowing there was a conflict. You cooperate when preserving the relationship is a priority. A **compromising conflict management style** occurs when both parties give up something of importance to arrive at a mutually agreeable solution to the conflict. This differs from the **collaborating conflict management style**, in which both parties work together to arrive at a solution without having to give up something of value.

When faced with conflict, your goal is to create a dialog with other involved parties in an effort to create a solution that is fair to all. This is called **negotiation**. Negotiation is not only a critical element of conflict management, it is also a healthy part of common workplace activities, including salary and job assignments. When negotiating, first prepare by identifying your optimal desired outcome for you and the other party. Also, identify the best and worst case scenario that can occur during your negotiations, including what outcome is unacceptable to you. After you have prepared for your negotiation, meet with the other party to share basic facts (i.e., each party's desired outcome) and agree on an objective. At that point, begin finding common ground and working toward an agreement.

Both sides can come to an agreement if both parties:

- Want to resolve an issue
- Agree on an objective
- Honestly communicate their case/situation
- Listen to the other side
- Work toward a common solution that is mutually beneficial

In working toward a successful negotiation, first and foremost, be ethical in your negotiations. The goal is to provide a win–win situation.

Also, practice good communication skills. Listen; do not interrupt or pass judgment until the other party has stated its case. Word choice matters. Consistently monitor less-than-positive relationships to ensure that your words are not fueling an already tense situation. As you learned in Chapter 9, observe the other party's body language through hand and arm gestures and body positioning. Attempt to identify whether the other party is willing to resolve the issue. Also evaluate the party's ability to make eye contact. Put aside personal feelings and focus on coming to a mutually agreeable solution.

You are displaying **passive behavior** if you consistently allow others to have their way. Although it is acceptable to use passive behavior at times, there are appropriate times to use assertive behavior. **Assertive behavior** is when you stand up for your rights without violating the rights of others. This is done by displaying confidence and not being ashamed to defend your position by sharing your concerns in an inoffensive manner. Professionals behave in an assertive manner, not in an aggressive manner. Individuals exhibiting **aggressive behavior** stand up for their rights in an offensive way that violates others' rights. Others do not have to be harmed or put down for you to be heard or to have your way. When aggressive behavior is exhibited, someone loses. Consistently treat others in a respectful, professional manner. If you are offended or see that someone else's rights are being violated, speak up. Stand up for your rights (or those of another) without harming others. Do not demean others as a form of retaliation when someone displays aggressive behavior toward you.

Copyright © 2016, 2013, 2011 by Pearson Education, Inc.

Talk It Out

What prevents individuals from being assertive?

Harassment

Offensive, humiliating, or intimidating behavior is called **harassment**. As discussed in Chapter 8, the human resource department is your advocate in cases of workplace harassment. There are various anti-discrimination laws that protect individuals from harassing behavior. One of the most common forms of workplace harassment is sexual harassment. The Equal Employment Opportunity Commission (EEOC) defines **sexual harassment** as unwanted advances of a sexual nature. The two types of sexual harassment are quid pro quo and hostile behavior. **Quid pro quo harassment** is behavior that is construed as payback for a sexual favor (e.g., you sleep with your boss and the boss gives you a raise in return). The EEOC states that quid pro quo harassment can include "verbal, visual or physical conduct of a sexual nature." **Hostile behavior harassment** includes any behavior by another employee that is of a sexual nature that you find offensive. The EEOC states this behavior can include "verbal slurs, physical contact, offensive photos, jokes, or any other offensive behavior of a sexual nature." Sexual harassment can occur between a man and a woman, a man and another man, or a woman and another woman. It does not limit itself to a boss and employee relationship.

In addition to a sexual harassment policy, companies should also have a policy regarding professional behavior. Each employee is entitled to be treated in a respectful and professional manner by coworkers. A policy regarding workplace behavior prevents incivility and communicates to all employees that unprofessional behavior will not be tolerated. You do not have to be friends with all of your coworkers, but you are required to respect colleagues and treat them in a professional manner. There is no place for rudeness in the workplace. You also do not have to like your coworkers, but they should not be aware of your negative feelings toward them. Mature adults treat all coworkers with courtesy and respect.

Exercise 12.2

What kind of behavior have you exhibited that could be construed as harassing? How can you change your behavior or attitude to ensure that no one is offended by or could misinterpret your behavior as inappropriate?

_____

_____

_____

All companies should have an anti-harassment policy. Employers need to provide training and a protocol for filing and investigating a complaint. If you are a victim of harassment, take the following steps:

1. If the behavior is offensive but relatively minor, tell the individual that his or her behavior is offensive and ask him or her to stop. Document the conversation in your personal notes. Include the date, the time, and any witnesses to the incident.

2. If the behavior continues or if the behavior is extremely inappropriate and/or outrageously offensive, immediately contact your supervisor or human resource department. Tell the person you contact what happened, that you are offended by the harassing behavior, and that you want to file harassment charges. Provide facts and names of anyone who witnessed the offensive behavior.

Once you have filed a complaint with your employer, he or she has a legal obligation to conduct a confidential investigation. Everyone is innocent until proven guilty. Do not feed the rumor mill with information regarding your complaint. Keep the issue confidential. Remain professional and reserve comments for the investigative interview. Document the dates and times of whom you speak with regarding the complaint, interviews, and comments made. When the investigation is complete, the supervisor or human resource department will render a decision. If you are not satisfied with the outcome, you have the right to file a complaint with your state's Department of Labor, Department of Fair Employment and Housing, or the Federal Equal Employment Opportunity Commission. It is unlawful for an employer to retaliate against or punish anyone who files a sexual harassment claim, even if the claim is found to be without merit. No one should be punished for filing a claim. However, employees should file only legitimate complaints.

Many companies have "zero tolerance policies," which means the company strictly enforces its anti-harassment policies. Harassment policies are extremely important. Behave appropriately, and do not exhibit behavior that could be offensive to others. This includes off-color jokes, inappropriate touching, inappropriate conversations, and suggestive attire. Many times, individuals think they are joking when in fact their behavior is offensive to others.

Web Search

Conduct a web search to identify an online sexual harassment quiz to gauge your understanding of appropriate workplace behavior.

Topic Situation

Katie, the mailroom clerk who works in Logan's office, returned from a vacation at the beach. When Logan asked Katie about her vacation, she told Logan and other coworkers all about her new tattoo. One of Logan's coworkers, Raj, asked where on her body she got her tattoo. Katie grinned at Raj and patted her chest. The next day, when Katie delivered the mail, Raj asked Katie when he would get to see her tattoo. Katie grinned and went on her way. For the next few days, Raj kept asking about the tattoo and Katie kept grinning and walking away. The following week, Raj was called into the boss's office. Logan later found out that Katie filed sexual harassment charges against Raj based on Raj's curiosity regarding Katie's new tattoo.

TOPIC RESPONSE

Who was right and wrong in the situation between Katie and Raj? Whose rights were violated? If you were Katie, should you have handled the situation differently? Why or why not? If you were Raj, what would you have done differently? Justify your answer.

Workplace Bullies

Another form of harassment is workplace bullying. Although you treat coworkers with respect, you may work with someone who is not reciprocating with the same respectful and professional behavior. An employee who behaves in an offensive, humiliating, or intimating manner is called a **workplace bully**. Workplace bullies are intentionally rude and unprofessional to coworkers. They seek ways to intimidate or belittle coworkers. Workplace bullying generally occurs among peers (the bully is on the same level as his or her target). Sometimes bullies publicly harass coworkers. Other times, they are discreet in their harassing tactics. Employees who are consistently rude, who bad-mouth other employees, or demonstrate intimidating behavior are displaying workplace incivility. Both bullying and incivility can result in a hostile work environment, which contributes to both an increase in stress-related performance issues and, worse yet, workplace violence. Workplace bullying and incivility are not only immature behaviors, they are unacceptable at work. If you experience workplace bullying:

- Do not retaliate with the same bad behavior. Remain calm and unemotional. The goal of a bully is to see that he or she has upset you. Though it is tempting to seek sympathy from coworkers, keep the issue confidential.
- Document dates, words used, and witnesses of the inappropriate behavior.
- Share the factual and documented information with your boss or human resource department and file a formal complaint.
- If you think your company has not appropriately resolved the issue in a reasonable time and manner, seek outside assistance. This assistance can come from a union, a counselor or mental health professional, a state or federal agency, or a private attorney.

Know Your Rights

Every employee has a right to work in an environment free from harassment, discrimination, and hostility. Your boss and human resource department cannot assist you in resolving conflict if they are not aware of the issue. Share your concerns and documentation regarding harassment, discrimination, and workplace incivility immediately with your superior. Prior to seeking outside assistance, exhaust all internal remedies (company resources that exist to take care of these issues). If you think you need outside assistance to preserve your rights, several state and federal resources are available to assist you. These include your state's Department of Fair Employment and Housing, the Federal Equal Employment Opportunity Commission, your State Personnel Board, the Department of Labor/Labor Commission, and the Department of Justice. These resources are available to act on your behalf and ensure you are being treated in a fair and non-discriminatory manner.

Resolving Conflict at Work

Several steps should be taken when attempting to resolve a conflict at work, including workplace harassment. They are presented in Figure 12.1. Whenever you are faced with a conflict, attempt to resolve the issue confidentially and as quickly as possible. Too often, individuals ignore a problem and hope it will go away. Unfortunately, these unresolved molehills frequently grow into giant, unresolved mountains. If you choose to utilize an accommodating or avoiding conflict resolution style, accept your decision to not bring the conflict to the attention of the offender and move on without holding a grudge.

If the conflict is negatively affecting either your or someone else's performance, inform your immediate supervisor. During this step, think like a boss and ask yourself if the matter is appropriate to be brought to the attention of a superior. You do not want to appear as a complainer. Document relevant information and conversations. If the problem continues and you are not satisfied with the way your immediate supervisor is handling the situation, contact your human resource manager. The human resource manager will review existing policies and your documentation. If the situation warrants, an investigation will be conducted. If you are not fully satisfied with the decision or handling of

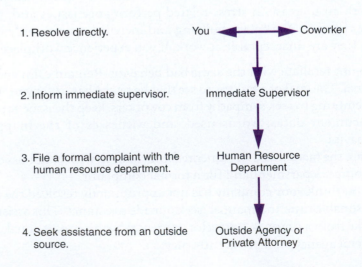

1. Resolve directly.   You ←——→ Coworker

2. Inform immediate supervisor.   Immediate Supervisor

3. File a formal complaint with the   Human Resource
   human resource department.   Department

4. Seek assistance from an outside   Outside Agency or
   source.   Private Attorney

Figure 12.1

Resolving Conflict

the situation, you have the right to seek assistance from a private attorney or an outside government agency identified earlier in this chapter. Prior to seeking assistance from any outside agency, attempt to resolve the problem within your organization's structure. If you have representation from a union, involve the union as early in the process as possible.

Resolving Conflict Under a Union Agreement

A union is a third-party organization that protects the rights of employees and represents employee interests to an employer. If you belong to a union and you have a conflict with your supervisor or any other member of management, refer to your union contract to identify what steps and rights are afforded you as a union member. Each workplace represented by a union has a **shop steward**, a coworker who is very familiar with the union contract and the procedures available to assist you in resolving a workplace conflict. A problem, unfair treatment, or conflict related to employment is called a **grievance**. Go to your shop steward and share your concern, along with any documentation or evidence you may have. If the conflict is valid and covered in your union agreement, the shop steward will meet with you and your supervisor in an attempt to resolve the issue. If the issue cannot be resolved at this level, a union representative will meet with the human resource department. The issue will continue to move up the chain of command until it is resolved. The formal steps taken in resolving a conflict between the union and an employer is called a **grievance procedure**. If you are represented by a union and your conflict is valid, your union representatives will assist you in protecting your rights through the grievance process. However, the union's purpose is not to shield you from punishment if you are guilty of wrongdoing. The purpose of the union is to enforce the union contract.

Workplace Violence

Unresolved conflict has the potential to escalate into workplace violence. According to the U.S. Department of Labor, workplace violence is the third leading cause of fatal workplace injuries. Workplace violence includes any type of harassing or harming behavior (verbal or physical) that occurs in the workplace. This violence can come from coworkers, a boss, a customer, or a family member. It is vitally important that you recognize the warning signs and take appropriate precautions to decrease the probability that you become a victim.

If you are a victim of harassment, seek assistance and report unprofessional behavior to your boss or to the human resource department before the behavior escalates to violence. Personal issues can also affect workplace performance. If you have a personal issue that you feel may negatively affect the workplace, share your concerns and seek confidential assistance from a coworker, your boss, or the human resource department as soon as possible. Some companies offer **employee assistance programs (EAPs)**. An EAP typically provides free and confidential psychological, financial, and legal advice. This benefit is generally extended to everyone who lives in your household. If you are experiencing a stressful situation at work or home, take advantage of this benefit. Even if your company does not offer an EAP, seek assistance from your human resource department as they may be able to help you identify an appropriate community resource.

TOPIC RESPONSE
Did Claudia handle the situation appropriately? Why or why not?

Topic Situation

Claudia shares a cubicle with a woman who is newly married. She appeared to be happily married and had told Claudia about the romantic dinners and gifts her new husband provided. One day, the woman showed up to work and kept to herself. Throughout the morning, Claudia found it strange that her coworker kept covering her face. Finally, as the lunch hour neared, Claudia asked the woman if everything was okay. The woman looked up at Claudia with a black eye and a bruised face and told Claudia that she was leaving her husband. She went on to tell Claudia that the husband had become increasingly jealous of any friendship the woman had with her male coworkers and had hit her the night before. Claudia asked her if she felt safe. The woman responded, "No." Claudia reminded the woman about the confidential and free EAP benefit their company offered. Claudia then assisted her in ensuring her safety and getting immediate help.

As exemplified in Claudia's experience, stress at home can affect performance at work. Many times, victims of domestic violence fail to seek assistance out of embarrassment or fear. Your employer will want to assist you. If you feel your conflict (either at work or at home) does not warrant professional assistance, find a friend with whom you can confidentially and neutrally discuss the issue. Take responsibility at work to ensure a safe working environment. Do not be afraid to ask for an escort to and from your car if you are working non-traditional work hours or if your car is parked in a remote location. Keep emergency phone numbers readily available and know where all the emergency exits are located. Report suspicious behaviors or situations that have the potential to become violent. It is much better to be safe than sorry.

Agree to Disagree

As we learned in this chapter, sometimes conflict cannot be avoided. In your efforts to work in harmony with coworkers, you will find that you will have coworkers whom you do not get along with or who may hurt you. As important as it is to apologize when we have harmed others, it is equally as important to forgive. Too often, coworkers have apologized and the harmed individual has failed to forgive. Forgiving does not mean that you have forgotten the hurt. It does mean that you will give the individual another opportunity to prove his or her apology was sincere by a change in behavior.

A mature coworker is always willing to forgive and not hold grudges. Those who hold grudges never forgave in the first place and may be seeking a means of retaliation. Doing so demonstrates immaturity. You do not have to like all of your colleagues, but you need to demonstrate professionalism and respect toward them. Conflict at work is inevitable. How you allow the conflict to affect your performance is your choice.

Think About It

Identify grudges you have held or people you need to forgive. Make a point of resolving one of those issues within the next week.

Workplace Dos and Don'ts

Do attempt to resolve a conflict as quickly as possible	*Don't* allow a small conflict to grow over time
Do utilize the appropriate conflict management style	*Don't* demonstrate aggressive behavior when standing up for your rights
Do know your rights regarding sexual harassment and discrimination issues	*Don't* utilize offensive language or hostile behavior
Do document any activity that you feel may escalate into potential problems	*Don't* retaliate when a workplace bully behaves inappropriately
Do agree to disagree when a conflict cannot be resolved	*Don't* hold grudges or behave in an immature manner

Concept Review and Application

You are a Successful Student if you:

- Name and describe the various conflict management styles and the appropriate time to use each one
- Write a plan detailing specific actions you will take to improve your personal negotiation skills
- Research and identify innovative methods companies are utilizing to prevent workplace bullying

Summary of Key Concepts

- A natural element of working with others is conflict.
- How you deal with conflict determines your maturity level and professionalism.
- Depending on the offense and workplace situation, there are several methods of dealing with conflict.

- Employees have a right to work in an environment free from harassment.

- Immediately report any harassing behavior.

- Attempt to resolve conflict at the lowest level, as soon as possible.

- Recognize the warning signs and take appropriate precautions to decrease the probability that you become a victim of workplace violence.

- It is considered immature to hold grudges. If you cannot resolve a conflict, sometimes it may be best to agree to disagree.

Self-Quiz MATCHING KEY TERMS

Match the key term to the definition using the identifying number.

Key Terms	Answer	Definitions
Accommodating conflict management style		1. Trying to make others do things your way
Aggressive behavior		2. Employees who are intentionally rude and unprofessional to coworkers
Assertive behavior		3. Standing up for your rights without violating the rights of others
Avoiding conflict management style		4. Unwanted advances of a sexual nature
Collaborating conflict management style		5. When you consistently allow others to have their way
Compromising conflict management style		6. A coworker who is very familiar with a union contract and can assist in resolving a workplace conflict
Conflict		7. Employee benefit that provides confidential psychological, financial, and legal advice
Employee assistance program (EAP)		8. Ignoring the offense to avoid conflict
Forcing conflict management style		9. Formal steps taken in resolving a conflict between the union and an employer
Grievance		10. Creating a dialog with other involved parties in an effort to create a solution that is fair to all
Grievance procedure		11. Behavior that is construed as payback for a sexual favor
Harassment		12. Both parties work together to come to a solution without having to give up something of value
Hostile behavior harassment		13. Allowing the other party to have his or her way
Negotiation		14. A problem, unfair treatment, or conflict related to employment

Key Terms	Answer	Definitions
Passive behavior		15. Any behavior by another employee that is of a sexual nature that an individual finds offensive
Quid pro quo harassment		16. Standing up for your rights in a way that violates others' rights in an offensive manner
Sexual harassment		17. A disagreement or tension between two or more parties
Shop steward		18. Offensive, humiliating, or intimidating behavior
Workplace bullies		19. When both parties give up something of importance to arrive at an agreement

Think Like a Boss

1. A fellow supervisor usually argues about an issue before arriving at a decision. Knowing this is typical of this person's behavior, how should you handle your next confrontation?

2. One of your employees tells you that another employee has been harassing him. What should you do?

Activities

Activity 12.1

Based on what you learned in this chapter, identify the appropriate response to these poor behaviors sometimes displayed by colleagues.

Poor Behavior	Appropriate Response
1. Uses foul language	
2. Steals company property	
3. Tells you to lie	
4. Verbally harasses other employees	
5. Takes all the credit for everyone else's work	

Activity 12.2

You have a coworker who believes his ideas are always the best and will not compromise. Applying each conflict management style, how would you respond?

Conflict Management Style	Your Response
Forcing	
Avoiding	
Accommodating	
Compromising	
Collaborating	

Activity 12.3

You believe you have worked hard this past year. Applying what you have learned in this chapter, how would you negotiate a raise or promotion?

_____

_____

Activity 12.4

Identify a time you felt you were harassed or had your rights violated. Based on the information you learned in this chapter, what should you have done differently?

_____

_____

What outside resource could/should you have contacted?

_____

_____

Activity 12.5

You are a member of a union and your boss is accusing you of wrongdoing and wants to meet with you. How should you respond?

_____

_____

_____

13

Job Search Skills

target • research • network

After studying these topics, you will benefit by:

- Conducting a job search in a targeted career, industry, and location
- Ensuring a professional online identity and protection of privacy
- Collecting items to be included in a job search portfolio
- Identifying references to be used in a job search
- Discovering sources for job leads
- Describing how networking is a powerful job search tool
- Explaining appropriate behaviors to utilize during the job search process
- Summarizing the importance of maintaining the right attitude during a job search

HOW DO YOU RATE?

Are you job search savvy?	True	False
1. It is best to attend a job fair prepared as if it is an interview.	☐	☐
2. It is acceptable to distribute personal business cards at social functions.	☐	☐
3. It is not necessary to share personal information such as a birthdate and Social Security number during a job search.	☐	☐
4. A job search portfolio is a foundation for the interview portfolio.	☐	☐
5. Most realistic job leads are found through networking.	☐	☐

▶ If you answered "true" to four or more of these questions, congratulations—you are well on your way to finding the job of your dreams. Knowing how the job search process works, creating a job search plan, and properly utilizing job search tools pave the way to job search success.

The Job Search

An effective job search is the key to finding a great job. A successful job search involves creating a plan, conducting research, and taking action. Doing so takes time, organization, communication, and professionalism (all key skills you have developed throughout this text). This chapter is designed to help you create a job search strategy. A successful job search strategy identifies what type of job you will be looking for, what tools and resources you will need, and how these tools and resources are best used. The job search is an integral part of your life plan as it brings life to your career goals. The ultimate goal of a job search is to secure an interview that paves the way toward obtaining the job of your dreams.

Choosing the Right Career

Creating a job search plan begins with choosing the right career. This involves **self-discovery**. Self-discovery is the process of identifying key interests and skills built on the career goals you wrote in Chapter 1. Knowing your key knowledge, skills, and abilities and linking these with your career goals will assist you in landing a job you will enjoy. This process includes identifying key interests and accomplishments from your work, educational, and personal experiences. A method for identifying key interests is creating an accomplishments worksheet, shown in Activity 13.1. The worksheet inventories skills you have acquired from work or non-work experience and serves as a tool to assist you in identifying the right career. Utilize power words when completing the accomplishments worksheet. **Power words** are action verbs that describe your accomplishments in a lively and specific way. These power words provide an excellent foundation when you create your personal commercial and build your résumé. Table 13.1 lists power words that will assist you in identifying your accomplishments.

Table 13.1	Power Words		
Adapted	Communicated	Instructed	Projected
Addressed	Coordinated	Installed	Recommended
Analyzed	Created	Introduced	Risked
Arranged	Determined	Investigated	Saved
Assisted	Developed	Learned	Staffed
Built	Earned	Located	Taught
Calculated	Established	Managed	Typed
Chaired	Financed	Motivated	Updated
Cleaned	Implemented	Organized	Won
Coached	Increased	Planned	Wrote

After you have completed your accomplishments worksheet, reread your responses. They will most likely reveal a targeted career of interest to you.

A second means of identifying key skills and jobs of interest is to take a formal career assessment to identify what careers best match your interests and abilities. Common career assessment tools include the Golden Personality Type Indicator, the Myers-Briggs Type Indicator, and the Strong Interest Inventory. Many college career centers offer these assessments, as do various online sources. Sponsored by the U.S. Department of Labor, an excellent online resource is the ONET Interest Profiler.

Once you have identified careers and specific jobs that are of interest to you, conduct a realistic job preview. A realistic job preview identifies day-to-day activities and common tasks that are performed and required for a specific job and also makes you fully aware of both the positive and negative aspects of a specific job. Identify standard qualifications and education requirements. For example, if you are a felon, you may not be allowed to work in some areas of health care and education. There are other careers that require a clean DMV record or credit history. Thoughtfully researching and understanding what is required to secure and succeed in a desired job early in the career exploration process will save time and money.

After identifying a target career, make your job search personal by creating a career summary statement. This statement provides focus for your job search by summarizing your career objective, knowledge, skills, abilities, and accomplishments. Your career summary statement will be used for networking purposes and will be a vital element of your personal commercial. Complete Activity 13.3 to create your career summary statement.

Industry Research

One step in a successful job search is research. When a job fits your personality and skills, you are more likely to succeed. A satisfying career comes from performing a job you enjoy and working at a company that reflects your values. In Chapter 1 you created both career and personal goals. Conducting industry research will provide career information to reinforce that you have made the right career decision to support your life plan. Identify industries that require the key skills you possess.

Your research will most likely reveal that there are more industries that require your key skills than you thought. The easiest way to conduct industry research is by using the ONET Database. This database of occupational information was developed for the U.S. Department of Labor and provides key information by job title. Match the key knowledge, skills, and abilities required for your target job with the knowledge, skills, and abilities you possess.

Once you have identified industries requiring your skills, begin identifying specific jobs in these industries. Note the different job opportunities and various job titles that exist within each industry. Being aware of the various job titles for which you qualify allows increased job search flexibility. After determining industries and job titles that fit your skills, identify the various environments available, including where the jobs are located and the specific work setting desired.

For example, if you finished college with a business degree, conduct research on industries that need employees with a business background, such as health

Web Search

Search and take the ONET Interest Profiler.

Talk It Out

What career area do you believe suits your skills and experiences?

Talk It Out

Discuss the difference between a job and an industry.

care, retailing, and manufacturing. Once an industry is identified, begin reviewing job titles that match the skills you have acquired in college, such as a financial analyst, general accountant, marketing assistant, or human resource generalist. After identifying specific job titles that match your skill sets, decide what type of work environment you desire. If you select health care, you may have the choice of working in a hospital, a clinic, or a private physician's office.

Conducting industry and work environment research will provide you information that will simplify your job search. Instead of sending out hundreds of résumés in hopes of securing any job, target companies that are a good match with your life plan, your skills, and your desired work environment.

The Targeted Job Search

After you have a clearly defined career summary and have identified jobs that suit your personal and career goals, it is time to begin a targeted job search. A **targeted job search** leads you through the process of discovering open positions for which you are qualified, in addition to identifying specific companies for which you would like to work.

Part of a job search is to determine in what city you want to work. If your job search is limited to your local area, you will be restricted to local employers. If you are willing to commute outside of your area, determine how far you are willing to commute (both directions) on a daily basis. If you wish to move out of the area, identify what locations are most appealing. Should you desire to move to a new location, consider the cost of living in your desired location. The **cost of living** is the average cost of basic necessities such as housing, food, and clothing. For example, it is much more expensive to live in Manhattan, New York, than it is to live in Cheyenne, Wyoming. Although a job in Manhattan may pay a lot more than a job in Cheyenne, living expenses typically justify the higher salary.

Exercise 13.1

Identify three companies/employers in your target location that may be of interest to you.

_____

_____

_____

Online Identity

Web Search

Conduct an Internet search on yourself.

With the popularity of social networking sites, your personal life has a greater chance of being exposed in the job search process. Ensure you have a favorable online identity. As explained in Chapter 10, an online identity is the image formed when someone is communicating with you and/or researching you through electronic venues.

Because the majority of information on the Internet is public information, an increasing number of employers are conducting web searches on potential employees to gain a better perspective of the applicant's values and lifestyle.

With today's overabundance of electronic social networking and information sites, personal blogs, and other file-sharing services, ensure that defamatory photos, writings, or other material will not be a barrier in your job search. When conducting an Internet search on yourself, remove any information that portrays you in a negative light. If you are actively involved in social networking sites, carefully evaluate any personal information such as photos and/or statements about you on these sites. If negative information is on your personal site, immediately remove it. If it is contained on sites of your friends, explain your job search plans and politely ask them to remove the potentially harmful information.

An additional step toward ensuring a clean online identify is to maintain a professional e-mail address. Sending a potential employer an e-mail from the address "prty2nite" is not the image you want to project. If necessary, establish a new e-mail address that utilizes some form of your name or initials to maintain a clean and professional online identity. Two final considerations in maintaining a professional online identity are, as mentioned in Chapters 9 and 10, the maintenance of a professional voice mail message and the avoidance of text slang in all written communication. Your job search strategy will involve extensive communication with employers and other individuals who will assist you with your job search. Interaction with these individuals needs to be professional.

> ### Think About It
> Does your social network site contain information that may hinder a friend's job search?

> ### Talk It Out
> What type of photos, writings, or materials do you think are inappropriate for a potential employer to see?

Job Search Portfolio

A **job search portfolio** is a collection of paperwork used to keep you organized and prepared while searching for a job. Some items from your job search portfolio will become a part of your interview portfolio, which you will create in Chapter 15. These items and their purpose will be discussed in this and the next two chapters.

When creating and managing your job search portfolio, it is best to have a binder with tabs to keep all paperwork organized and protected. When you begin collecting items for your portfolio, have the original and at least two copies of each item available at all times. Copies will be transferred to your interview portfolio when needed. Original documents should not be removed from your job search portfolio. Place original documents in plastic notebook protectors; do not punch holes in original documents.

Because many of today's job searches occur over the Internet, it is also recommended you create an electronic job search portfolio. An **electronic job search portfolio (e-portfolio)** is a computerized folder that contains electronic copies of all job search documents. For your electronic job search file, scan copies of all documents you will be keeping in your hard-copy portfolio. Save these scanned files in portable document format (.pdf) to ensure the documents are able to be easily retrieved by the receiver. Keep your electronic job search portfolio in an electronic folder that is both easy to retrieve and access. When sharing job search files with potential employers and others, these electronic documents will be sent as attachments. The electronic job search portfolio may be very useful in your job search should an employer request documentation, as you will have immediate access to this information through a mobile electronic device.

You may be sharing some of these items with employers, so when creating documents, proofread and ensure they are professional and error-free. Table 13.2 is a list of items to include in your job search portfolio.

Table 13.2   Job Search Portfolio

Item	Description
Awards	Documents that demonstrate proficiency in specific skills
Certificates	Documents that demonstrate proficiency in specific skills
Completed generic application	Generic job application that makes information readily available
Copy of ID and/or driver's license	A valid ID and proof of ability to drive (if driving is a job requirement)
Copy of recent DMV record (if relevant to your career)	Used to ensure a safe driving record
Cover letter	Introduces a résumé
Current state licenses (if relevant to your career)	Documents that verify the ability to practice certain professions
Job search notebook	Three ring binder to store all job search documents secure in one place
Letters of recommendation	A written professional reference to verify work experience and character
Network list	A list of professional relationships used for job contacts
Pen	Used for tracking and keeping notes
Performance appraisals from previous jobs	Proof of positive work performance
Personal business cards	Cards with personal contact information used to share for job leads
Personal commercial	Statement that assists with interview
Plastic inserts	Use for originals (do not place holes in originals)
Reference list	A list of individuals who will provide a professional reference
Résumé	A formal profile that is presented to potential employers
Small calendar	Used to track dates
Small note pad	Use for notes
Thank you notes with a draft message	Use for post-interview follow up
Transcripts	Documents that verify education (have both official and copies available; sealed transcripts may be required)
Work samples	Documents that demonstrate proficiency in specific skills

Many careers that involve driving require a copy of your driving history. This information is secured by contacting your local Department of Motor Vehicles (DMV). If you have a poor driving record, check with your local DMV to identify how long this history stays on your record. Those with a blemished driving history may have a tougher time securing a job in a field that involves driving. When sharing your DMV record, as with all other portfolio items, provide only a copy (unless otherwise required) and maintain the original in your job search portfolio.

Employment Applications

Keep a completed generic employment application in your job search portfolio so you have required information readily available. If you have a smartphone, simply take a photo of the completed application and store this information on your device for quick and easy retrieval. To protect yourself from potential identity theft, do not list your Social Security number or birthdate on any employment application, if possible. On some electronic applications, this information is a required field. If such is the case, ensure you are utilizing a secured site.

An employment application is a legal document. When completing the application, read the fine print prior to signing the document. Commonly, at the end of the application, there will be a statement that grants the potential employer permission to conduct reference and various background checks, including a credit check, if the information is relevant to the job for which you are applying. Fully understand why this background information is necessary and how it will be used in the hiring process. If you do not fully understand the statements on the application, get clarification prior to signing the application.

It is common for employers to request that the applicant complete an employment application and submit this document along with the résumé package. If you submitted only a cover letter and résumé, you may be asked to complete an application after you have been interviewed. Many employment applications can now only be completed and submitted through a kiosk located at a worksite or downloaded, completed, and submitted directly through a company website. A typed employment application is best. If typing is not possible, complete the application by printing neatly in black ink. In some instances, after you have completed an online application, you may be asked to take a pre-employment test as part of the application process. An in-depth discussion on pre-employment tests is presented in Chapter 15.

Personal References and Recommendations

Create a list of professional references that a potential employer can contact to verify your work experience and personal character. While a professional reference list is part of your résumé package, references are not to be included on your résumé; list them on a separate page. Do not send your reference list with your résumé unless it is requested by the employer. However, have a copy available to share if the employer requests references during the interview. References can be from past or present employers and supervisors, coworkers, instructors, or from a representative of an organization with which you have volunteered. Do not use relatives, friends, or religious leaders unless you have worked or volunteered with or for them.

Prior to including individuals on your reference list, ask each person if he or she is willing to provide a reference. When asking someone to serve as a reference, share with them details on your job search, including your career goal, target employers/job, and the status of your search. Make certain each person on this list will provide a positive reference. Have at least three names to submit as references. Include each reference's name, relationship, contact phone number, e-mail address, and business mailing address. An example of a professional reference list is provided as part of a sample résumé package in Chapter 14, figure 14-7.

In addition to reputable references, it is wise to have at least three **letters of recommendation**. A letter of recommendation is a written testimony from another person that states that you are credible. Begin collecting letters of recommendation before you actively engage in a job search. Strong letters of recommendation reflect current job skills, accomplishments, and positive human relations skills and should be no older than one year. Letters of recommendation can be from past or present employers, coworkers, instructors, or someone you worked for as an intern or volunteer. It is common and acceptable to have someone write a formal letter of recommendation and serve as a personal reference. When asking for a letter of recommendation, provide details about the job skills, accomplishments, or human relations skills you feel they observed. This information will be valuable when they write your letter. Provide your reference a minimum of two weeks to write your letter. If possible, have the letter written on company or personal letterhead. Also ensure the letter is dated and signed. After receiving your letter, immediately send your reference a note or small gift of thanks for their effort.

Ask your reference to include the following information in the reference letter:

- Connection to you, why he or she knows you are qualified, and why he or she is providing a reference letter
- Information about your skills and how they would benefit the company (provide information and/or a copy of the job qualifications if applicable)
- A summary of why he or she is recommending you
- Reference contact information

In addition to routinely updating your résumé, keep your reference list updated. References listed should be relevant to your career goal. Occasionally contact your references to verify that they are still willing to serve as references. Keep these individuals current on your job search status and career goals.

Think About It

Other than an instructor, identify someone from whom you can secure a personal letter of recommendation.

Sources of Job Leads

Your search for job leads is a job in itself. Potential employers will not seek you. You must actively seek them. Fortunately, there are many available sources for job leads. The most obvious job lead is directly from a targeted company. View the company website or personally visit the target company's human resource department for current job announcements. It is also common for employers to post job announcements on social networking and job search websites. If you do not have a targeted company but have a location where you would like to work, conduct an Internet search using the target city and position as key search words. Search associations in your targeted industry to view industry-specific online job banks. Check online message boards and popular job search sites. Maintain a record of the sites you are utilizing for your job search and monitor activity. Many larger cities and counties offer one-stop centers for job seekers. These government-funded agencies provide job-seeker assistance and serve as a link between job seekers and local employers. Other job sources include job fairs, newspaper advertisements, industry journals, and current employees who work in your targeted industry and/or company.

Do not rely solely on posted job positions. Many jobs are unsolicited (not advertised to the general public). The way to become aware of these unsolicited jobs is to use your professional network. A professional network is a group of relationships that are established primarily for business purposes. Use your network to identify individuals who either are or know current employees who work in your targeted industry and/or company. Inform network members of your desire for a job and ask for potential job leads. The larger your professional network, the more you will become aware of unannounced job leads. A discussion on how to create and utilize a professional network is presented in the next section.

Treat all face-to-face job search situations—including distributing your résumé, meeting a potential network contact, attending a job fair, or visiting a company to identify open positions—as if you are going to an interview. Dress professionally, go alone, display confidence, keep easy access to your online portfolio, and bring your interview portfolio. The interview portfolio will be discussed in detail in Chapter 15. In networking situations where there are many job seekers, such as a job fair, be polite and professional in your interactions with everyone. Do not interrupt or be rude to other job seekers. Take the lead in introducing yourself to company representatives. Sell your skills and confidently ask the company representative if he or she has an open position requiring your skills. The goal in such a situation is to favorably stand out from the crowd, share your résumé, and arrange an interview. There are situations where applicants are invited to "on the spot" interviews. Your professional appearance, preparation, and interview portfolio will demonstrate that you are a serious candidate for the job. Dressing casually and/or having a child or friend in tow will communicate unprofessionalism to a potential employer.

If you are unable to identify a job lead, send an unsolicited cover letter and résumé to your target company either electronically or through traditional mail. When sending an unsolicited résumé, send two copies: one to the human resource manager, and the other to the manager of your target job. Prior to sending your résumé, call or research the company to ensure you have identified the correct spelling and gender for the individuals to whom you will be sending your résumé. Sending two résumés to the same company increases the opportunity of securing an interview. The targeted department manager will most likely read and file your résumé for future reference. The human resource manager will also review your résumé and may identify other jobs for which you are qualified.

An internship is an excellent method of enhancing your job skills and a venue to expand your professional network. An internship is a paid or unpaid method of on-the-job training and can prove to a potential employer that you can handle the demands of the job. Many colleges provide assistance in securing an intern position. Approach an internship as an ongoing interview, as your manager and the individual with whom you will be working are evaluating you on your performance and professional behavior. Throughout your internship, demonstrate value to the company by being responsible. Also, use your internship as a means of increasing your professional network. At the end of a successful internship, it is appropriate to request a letter of recommendation from the employer and/or coworkers.

Professional Networking

According to the Bureau of Labor Statistics, over 70 percent of jobs are found through networking, although many estimate that number to be higher. Therefore, establishing and maintaining a professional network is important for career success. A **professional network** is a group of relationships that are established primarily for business purposes. A professional network is created through **networking**, the act of creating professional relationships. Think of networking as a connection device. The purpose of creating a professional network is to have a resource of individuals whom you can call on for professional

> **Talk It Out**
>
> What job fair behaviors demonstrate to a recruiter that you are professional?

> **Talk It Out**
>
> How can you specifically use an internship to increase your professional network?

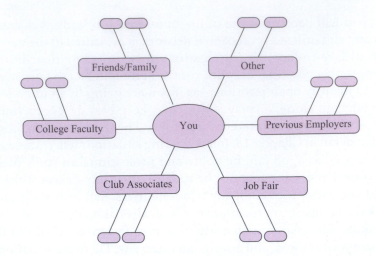

Figure 13.1

Professional Network

assistance and/or advice. The intent of this discussion is to utilize a professional network for job search purposes; however, a professional network is also a useful tool for collaborating and assisting others, as discussed in Chapter 6.

Creating a professional network involves meeting individuals and communicating a need. In the case of a job search, you inform one person that you are looking for a job. Ideally, that individual will inform others, then those people inform others, and soon you have many people who know that you are searching for a job. Should any of these individuals know of a potential job, they will feed the information back to you. Figure 13.1 provides a visual of a professional network.

Professional networking is necessary throughout a job search. There are two primary forms of networking. The first form is the traditional method, which involves face-to-face interaction. The second method utilizes social media. Traditional networking involves interacting with and meeting as many people as possible who work or know someone who works in your targeted industry. There are many formal networking opportunities for job seekers, including attending association meetings, service clubs, and conferences. Additionally, many college career centers provide networking events for students to interact with local employers. Job fairs, volunteer fairs, and trade shows are also excellent venues for professional networking. The key to successful networking is to create a network before you need one. Doing so provides time to develop networking skills, increase your confidence, and identify which networking venues work best. Many college recruiters enjoy meeting students a year prior to graduation. Students who network early in their job search convey organization, planning, and strategic skills, which are behaviors highly desired by employers.

Create and maintain a **network list**, an easily accessible list of all network contacts' names, industries, e-mail and traditional addresses, and phone numbers. Provide contacts that are actively assisting with your job search a copy of your most current résumé and keep them updated on your search. When communicating with members of your network, be sensitive to their time. Do not annoy them or be inconsiderate in your interactions. Ensure that your network contact list is current. Find and consistently utilize a database system that is convenient for you. Most individuals use an electronic database, although some still prefer a traditional address book. When you secure a job, immediately remove job search postings and inform members of your network who were actively assisting you with your job search.

Talk It Out

Which type of network list (traditional or electronic) works best for your needs and why?

Almost every person you know may be a part of your network, including coworkers, supervisors, instructors, family, and friends.

Topic Situation

Tran has been working as a mechanic for a year. During this time, Tran has acquired new skills, learned the latest technologies related to auto mechanics, and earned a degree. He begins telling supervisors and coworkers what skills, education and technologies he has learned over the last year and shares career goals. Tran then tells family members and friends the same information.

TOPIC RESPONSE
Has Tran created a professional network? Why or why not? How can Tran enhance this network?

When engaging in traditional networking, remember that the success of one's networking attempts begin with a positive attitude. Review what you have learned in previous chapters of this text regarding professional behavior. First and foremost, believe in yourself and know what key skills and abilities you have to offer. Be confident and willing to approach and initiate an introduction with strangers. People are drawn to positive people. When engaging in face-to-face networking, dress professionally—first impressions matter. A useful networking and introduction tool is a personal business card. This small card contains contact information, including your name, mailing and e-mail addresses, and phone number. It is a good practice to share your personal business card with anyone you meet, especially in networking, informational interview, and mentoring encounters. Doing so makes it easier for a new acquaintance to remember and contact you in the future. Personal business cards are inexpensive and valuable networking tools. When designing a personal business card, ensure it contains all relevant contact information and reflects a professional image. Use an easy-to-read font style and avoid fancy graphics, pictures, or too many words. Simple is better.

Have your personal business card and a career summary statement of your key skills and abilities to informally share with anyone you meet. When you share a business card, ask for one in return so you can follow up and add these individuals to your network list. Practice the art of introducing yourself in a positive and mature manner, beginning with a professional handshake. Listen carefully to the name of the individual you are meeting. After the handshake, exchange business cards. Focus the conversation on the other person. Prior to telling your new contact about you and your job search, get the individual to talk about where he or she works, what he or she does, and what he or she enjoys about the job. Use this time to build rapport. At the appropriate time, tell the individual about you and your job search. As your conversation continues, watch for body-language cues. If the person is engaged, he or she will make direct eye contact and turn his or her body toward you. When the individual no longer wishes to visit, he or she will most likely look away and/or turn his or her body away from you. Utilize these cues to either continue the visit or politely thank him or her for their time and end the conversation.

Though it is common to have food and beverages available at formal networking events, it is best to refrain from eating and drinking until you have met your desired network contacts. Practice proper etiquette by not overindulging in food. You are in attendance to meet people, not to eat. Refrain from drinking alcohol.

Within 24 hours of creating a new contact, follow up with a brief message sharing with the individual that it was a pleasure meeting him or her.

Exercise 13.2

Write a brief message to someone you have recently met and would like to include in your network list.

_____

_____

_____

Additional network opportunities include volunteering for community organization, performing internships, and participating in work experience programs. A U.S. Census study found that volunteers have a 27 percent better chance of finding a job than those who do not volunteer. Volunteering provides opportunities to meet people in different organizations and learn about new positions. As mentioned in Chapter 11, volunteering is an excellent venue to develop leadership and team-building skills, network, and, most importantly, give back to your community. In addition to contacting a company directly for an internship or work experience opportunity, utilize your campus career resource center. Join clubs and professional organizations and actively participate. Attend professional workshops, conferences, and seminars to meet people from corporations that are in your targeted industry. Use the opportunity as a means of enhancing your skills and job search network. When performing these jobs, treat them as if you are in a paid job. Dress appropriately, use proper communication, perform all work in a quality manner, don't complain, and consistently behave in a professional manner. Be social, but avoid becoming personally involved with others. Get to know the employees and the company/organizational culture. Show your willingness to help solve problems and learn new skills.

Current technologies now provide ample social media outlets to not only post your résumé to targeted industries, but also to create an electronic network. Popular online professional networking venues include LinkedIn and Facebook. These venues offer special support services for job seekers, as do many industry-specific online networking sites. When utilizing a social networking site for your job search, ensure that your career information is current and consistent with your résumé. When there is an opportunity to post a professional photo of yourself, do so, as social media is a visual venue. It is also becoming increasingly popular to create and post a job video. The purpose of a job video is to utilize visual media to sell your skills. Dress professionally, practice and review what you will communicate (including body language), and make your video unique. After completing your video, solicit input from a trusted friend, and only post the video when you are certain it represents you in a positive and professional manner. If you are not comfortable on video, consider a blog or podcast as a means of utilizing social media tools. Be cautious when sharing personal information over social media venues by only posting on respected sites.

Earlier in this section, college career centers were suggested as a means of creating and/or expanding a professional network. Today's college career centers provide a tremendous amount of contemporary job search resources that reach beyond a traditional job fair. They provide job counseling services that include practice interviews, career assessments, workshops, networking events, and online job and interview information specific to your school and geographic location. It is common for these centers to tweet or text job postings to qualified students.

Talk It Out

What is appropriate and inappropriate information to share with an online network?

Think About It

Do you know where your career center is located? Have you taken advantage of the resources they have to offer?

Another means of expanding your professional network is to conduct an informational interview. An **informational interview** is when a job seeker meets with a business professional to learn about a specific career, company, or industry, not to ask for a job. Politely request an appointment, but do not attach your résumé. During an informational interview, ask the business professional questions about targeted careers, hiring, and the culture of the company. The process of meeting and talking with business professionals increases your professional network. When conducting an informational interview, explain who you are and the purpose of the interview. Inform the contact that the interview is only for career information gathering. Ask if you can have a few minutes of their time, usually about 20 minutes, to discuss their career and organization and to answer career-specific questions. Prior to the interview, research the company and prepare questions. During the interview use only the questions you prepared and respect time limits. If you conduct your informational interview in person, dress professionally and bring your career portfolio in case the interviewee requests a copy of your résumé. After the informational interview, promptly send a thank-you note.

Exercise 13.3

Identify appropriate questions to ask during an informational interview.

_____

_____

_____

Networking for both business and job search purposes is work, but the effort reaps tremendous benefits if done appropriately. Every few months, review your professional network list. If there is someone with whom you have not recently connected, contact him or her to say hello, share industry-specific information you think may be of value to them, or simply keep him or her updated on your career and growth plans. When networking, do not be afraid to ask your contact for additional contacts that might be able to assist you. Networking involves giving and taking. If you read an industry-related article, attend a conference, or are working on a project that may interest someone in your network, share the information and demonstrate how you can be of value to them.

Protecting Your Privacy

The job search process involves sharing personal information. Be cautious and share only personal information with reputable sources or you may become a target for identify theft. If you are applying for a job and have never heard of the employer, conduct research to verify that the employer is legitimate. As stated earlier, whenever possible, do not share your birth date or Social Security number with anyone, including employers, until after you are a finalist for a job.

TOPIC RESPONSE
What should Ravyn do?

Gene's friend Ravyn was looking for a job. Ravyn found a job on an online classified job site that sounded legitimate. The employer asked that Ravyn submit a résumé online. Within a few days after sharing her résumé, Ravyn received an e-mail telling her that she was a finalist for the job. The only step left in the process was for her to forward a copy of her credit report. Although Ravyn was desperate for a job, she thought this was a little strange, so she asked Gene what he thought of the situation. Gene helped Ravyn conduct an Internet search on the company, but could not find any evidence that the company even existed.

Keeping the Right Attitude

Throughout this text, you have learned how to be successful in the workplace. The importance of maintaining a positive attitude throughout your career cannot be stressed enough. This also holds true during your job search. The job search process is a lot of work and is often stressful and frustrating. Do not be discouraged if you do not get an interview or job offer on your first try. In tight job markets, it may take many interviews before receiving a job offer. Follow these tips to maintain a healthy attitude during this time of transition:

1. *Stay positive.* Start each day with a positive affirmation. Speaking aloud, tell yourself that you are a talented and great person who deserves a good job (and believe what you say). Your attitude is reflected in your actions. If you allow negativity to influence your job search, you will be at a disadvantage.

2. *Stay active.* Create a daily and weekly "to do" list. Every day, check the websites of targeted industries, associations, and companies in addition to relevant job sites. Schedule time for industry and company research. A job search is a job in itself. You do not want to be an unproductive employee in the workplace, so begin creating good work habits now by making the most of your time in a job search.

3. *Keep learning.* Use job search down time to learn or develop a skill. Identify a skill that will assist you when you are offered a job. Finances do not have to be a barrier to learning new skills. There are many free tutorials available on the Internet. Topics to consider include computer skills, writing skills, or any skill specific to your target industry.

4. *Stay connected.* Participate in networking on a weekly basis. Although it is natural to not want to socialize with others when discouraged, the job search period is the time when you most need to be in the presence of others. Keep your current network updated on your job search; identify valid reasons to communicate with your network. Consistently work on expanding your network by attending association meetings and events, volunteering, and scheduling informational interviews. Plan at least one meeting and/or activity each day. As opposed to sitting around the house waiting for the phone to ring, dressing professionally and networking every day will contribute to maintaining a positive outlook.

5. *Stay focused.* During this time of transition, manage your professional job search, your personal health, and your environment. Maintain an up-to-date calendar with scheduled follow-up activities relating to your job search. Because a job search is a stressful experience, practice healthy stress management techniques, including a proper diet, regular exercise, and positive self-talk. Invest a portion of your time in something of interest other than your job search. Consider volunteering for an organization of special interest to you. Doing so will provide a mental break, provide possible new network contacts, and provide you the satisfaction of helping others. Make wise choices regarding personal finances. Be cautious and conservative with your money. Make thoughtful purchases and avoid emotional spending. Finally, surround yourself with individuals who are positive and supportive of you and your efforts.

If you are currently working and are looking for a new job, keep your job search confidential unless you have completed training or education that will qualify you for an internal promotion. When in good standing with your supervisor, let him or her know you are looking for a new job and briefly explain why. Do not quit your current job before accepting a new job. Also, do not bad-mouth your company or anyone who works for your current or former employer(s).

MyStudentSuccessLab	Please visit **MyStudentSuccessLab**: Anderson|Bolt, Professionalism Skills for Workplace Success, 4/e for additional activities, resources, and outcomes assessments.

Workplace Dos and Don'ts

Do keep your original job search documents in a portfolio	*Don't* give employers your original documents and expect them to be returned to you
Do keep a network list and keep the people on your list updated	*Don't* be annoying or inconsiderate of your network contacts' time
Do realize that a targeted job search takes time	*Don't* get discouraged if you do not get an interview or job offer on your first try
Do explore various sources of job leads, including your personal network, the Internet, and industry journals	*Don't* limit your job leads to one source

Concept Review and Application

You are a Successful Student if you:

- Conduct a self-discovery assessment to identify the right career
- Create a formal network list
- Create a professional reference list

Summary of Key Concepts

- The career objective or personal profile is a brief statement that sells your key skills and relates to your self-discovery.

- A targeted job search leads you through the process of identifying open positions for which you are qualified, in addition to identifying companies for which you would like to work.

- Ensure you have a professional electronic image while job searching.

- Professional networking is the act of creating professional relationships.

- In addition to people you already know, develop additional network contacts through various sources of job leads.

- Creating and maintaining a job search portfolio will keep you organized and prepared during the job search process.

- Create a list of professional references for employers.

Self-Quiz MATCHING KEY TERMS

Match the key term to the definition using the identifying number.

Key Terms	Answer	Definitions
Cost of living		1. When a job seeker meets with a business professional to learn about a specific career, company, or industry
Electronic job search portfolio		2. The process of identifying key interests and skills built on career goals
Informational interviews		3. Average cost of basic necessities such as housing, food, and clothing for a specific geographic area
Job search portfolio		4. A written testimony from another person that states that a job candidate is credible
Letter of recommendation		5. The act of creating professional relationships
Network list		6. Action verbs that describe your accomplishments in a lively and specific way
Networking		7. Process of discovering positions for which you are qualified, in addition to identifying specific companies for which you would like to work

Key Terms	Answer	Definitions
Power words		8. A computerized folder that contains electronic copies of all job search documents
Professional network		9. An easily accessible list of all professional network contacts' names, industries, addresses, and phone numbers
Self-discovery		10. A collection of paperwork needed for a job search
Targeted job search		11. A group of relationships that are established primarily for business purposes

Think Like a Boss

1. What information would you supply to a job seeker during an informational interview with you?

2. If you discovered that one of your top interview candidates had an unprofessional website, what would you do?

Activities

Activity 13.1

Complete the following accomplishments worksheet. Use power words to answer each question. Whenever possible, quantify your answers by documenting how many, how often, and how much. Include education as well as non-work experience such as volunteerism.

Question	Response
1. Of what career-related activity are you most proud?	
2. Name a work- or school-related achievement.	
3. List major work and career-building tasks you have performed.	
4. What results have you produced from the tasks performed? (Include samples for your job search portfolio.)	
5. List three completed projects that demonstrate your ability to produce results.	
6. Provide a specific example of how you have successfully worked with others.	
7. What other life accomplishments make you proud?	

8. List extracurricular activities and volunteer work you have been involved with.	
9. List special skills or foreign languages you speak or write.	
10. What areas of interest do you have?	

Activity 13.2

Using ONET or other Internet resources, identify three specific job titles that match your career goals and current qualifications.

1. _____

2. _____

3. _____

Activity 13.3

Utilizing information from your accomplishments worksheet, career assessment, and realistic job preview, complete the following table to create a career summary statement.

	Example	Key Message
Target Job	Entry-level event planner	
Primary Skills	Organized, creative, attention to detail	
Qualifications	Marketing and business courses; customer service experience; bilingual (Spanish)	
Career Summary Statement	Organized, creative individual seeking entry-level event planner position. Bilingual (Spanish) with experience in customer service and successful completion of courses in marketing and general business.	

Based on the information from your accomplishments worksheet and your interest profiler results, assess whether your target job supports the life plan you created in Chapter 1. If not, what modifications need to be made to your personal, educational, and/or career goals?

_____

_____

_____

Activity 13.4

Create a job search portfolio by compiling the items from Table 13.2 and placing them in a binder.

Activity 13.5

Secure and complete a blank job application, with the exception of your signature. Add this document to your job search portfolio.

Activity 13.6

Name three issues to consider when identifying appropriate references.

_____

_____

_____

Activity 13.7

Complete the following reference list. This information will become part of your job search portfolio and the résumé package you will create in chapter 14.

	Your Name: **Address:** **City, Zip:** **Contact Phone:** **E-mail:**
Name	
Employer/Relationship	
Phone	
E-mail	
Address	
Name	
Employer/Relationship	
Phone	
E-mail	
Address	
Name	
Employer/Relationship	
Phone	
E-mail	
Address	

Activity 13.8

Use the following table to create a networking list (this document will become part of your job search portfolio).

Network List				
Name	**Address**	**Phone No.**	**E-Mail Address**	**Last Date of Contact**

Activity 13.9

Design a personal business card (this card will become part of your job search portfolio).

Activity 13.10

Create a five question informational interview sheet

1.	
2.	
3.	
4.	
5.	

14

Résumé Package

accurate • appealing • effective

After studying these topics, you will benefit by:

- Building a powerful résumé package
- Communicating a clear career objective/personal profile
- Utilizing power words and quantifiable outcomes to reflect personal accomplishments and experiences
- Identifying methods for effectively sharing a résumé
- Developing a cover letter
- Integrating methods to tailor the résumé package for target industries and employers
- Addressing special circumstances and time gaps

HOW DO YOU RATE?

Do you have résumé expertise?	True	False
1. Paper résumés are no longer necessary in today's electronic age.	☐	☐
2. Career objectives are used on all résumés.	☐	☐
3. Unique skills such as being bilingual or serving in the military can lead to discrimination and should not be listed on résumés.	☐	☐
4. Using a word processing résumé template is best when creating a résumé.	☐	☐
5. Those with a job gap on their résumé should make up a job to fill in the gap.	☐	☐

▶ If you answered "true" to at least two questions, use the information and tools in this chapter to improve your chances of creating and utilizing a winning résumé.

Building Your Résumé Package

Before employers decide to meet you, they first view your application materials. Traditionally, the set of application materials is referred to as a résumé package, which includes a résumé and a cover letter. Although many employers only allow online applications, which may not require a résumé and/or cover letter, every job seeker needs to create a formal résumé and cover letter. This information is used as a foundation for information an employer will require in the application process. The creation of a quality résumé package provides a job seeker with personalized, concise, and accurate information that can be used in any job search situation.

Your résumé package needs to efficiently and effectively sell your skills and communicate how your attributes are unique compared to all other candidates vying for your target job. A **résumé** is a formal written profile that presents a person's knowledge, skills, and abilities to potential employers. Your résumé is an important job search tool that should be created in advance of a job search and continually updated throughout your career. Even if you are not currently searching for a new job, a time will come when a current résumé is needed. Do not wait until that time to create or update your résumé. Continually add new job skills, accomplishments, and experiences to your résumé.

When you begin to create your résumé, you will quickly discover that there are various types of résumés and résumé formats. You may also receive conflicting advice as to how the perfect résumé should look and what it should include. The appropriate type of résumé used depends on your work experience, education, and other factors. A well-written résumé makes it easy for potential employers to quickly and easily identify your skills and qualifications that make you the right choice for the job.

This chapter will present the tools for creating a professional résumé and cover letter. As you go through the process of constructing your résumé package, make every word sell your skills and career accomplishments. Your résumé package represents you. Therefore, be honest with the information you provide and display character by not lying or embellishing the truth. There are five steps toward building a winning résumé:

1. Gathering Information
2. Creating an Information Heading and Utilizing Proper Layout
3. Writing a Skills Summary or Personal Profile
4. Inserting Skills, Accomplishments, and Experience
5. Reviewing the Completed Résumé

Step One: Gathering Information

The first step in building a résumé is to create a draft document with key headings. This involves collecting and merging all relevant information into one document. Begin identifying and listing the following information into an electronic document:

1. *Education.* List schools, degrees, certificates, credentials, GPA, licenses, and other relevant education-related information, including military experience. Include dates with each entry.

2. *Skills.* List all skills you possess and identified from the completed accomplishments worksheet (Activity 13.1) in Chapter 13 .

3. *Employment.* Starting with the most recent job, list the employer, start and end dates of employment (month and year), job title, and responsibilities.

4. *Languages.* List foreign languages, fluency levels, and if you can read, speak, and/or write the foreign language.

5. *Honors and awards.* List any honors and awards you have received at school, work, or from the community.

6. *Professional/community involvement.* List volunteer work and community service projects. Include any leadership role you took in these activities.

Note that when compiling information to include in your résumé, there is no personal information listed. Personal information such as birthdate, Social Security number, marital/child status, ethnicity, or religion should not be included on a résumé. It is also inappropriate to list hobbies or include photographs. There are laws that protect employees from discrimination in hiring and advancement in the workplace, and employers should not be aware of personal information unless it is relevant to the job for which you are applying. Additional information regarding this subject is presented in Chapter 15. Older job seekers should not list the date of graduation on a résumé, as it could be used for age discrimination.

Step Two: Creating an Information Heading and Utilizing Proper Layout

The second step in developing a successful résumé is to begin your electronic document. This includes listing personal contact information and identifying and arranging your information in the proper résumé layout. The top of your résumé is called the **information heading**. An information heading contains relevant contact information, including name, mailing address (city, state, ZIP code), contact phone, and e-mail address. Use your complete and formal name, including a middle initial if you have one. When listing your e-mail address, remove the hyperlink. If your current e-mail address is unprofessional, secure an address that is professional. Include only one contact phone number that is active with a professional voice mail message. Review the information heading for completeness, proper grammar and spelling, and accuracy. Spell out the names of streets. When using abbreviations, check for appropriate format, capitalization, and punctuation.

Once you have created your information heading, lay out your résumé. If you are at the start of your career and/or do not have extensive work experience, create a résumé using the **functional résumé layout**. This layout is used to emphasize relevant skills when you lack related work experience. A functional résumé focuses on skills and education. When writing a functional résumé, the first section contains the skills summary statement that you will create in step three. Immediately below the skills summary you will list relevant skills and education. Include your high school in the education section only if you have not yet graduated from college. Finally, list your work experience. Most functional

Functional Résumé Layout, see Figure 14.1 on page 218

Functional Résumé Example without Career Work Experience, see Figure 14.2 on page 219

Functional Résumé Example with Career Work Experience, see Figure 14.3 on page 220

Advanced Skill Set Résumé Layout, see Figure 14.4 on page 221

Advanced Skill Set Résumé Example, see Figure 14.5 on pages 222–223

Talk It Out

Which résumé layout is best for your situation? Why?

résumés are only one page in length. Refer to Figure 14.1 for the functional résumé layout and Figures 14.2 and 14.3 for examples of a functional résumé with and without career-related work experience.

Those with extensive career experience should use an advanced skill set résumé layout. An **advanced skill set layout** best highlights, communicates, and sells specific job skills and work accomplishments. In the advanced skill set layout, the skills summary is replaced with a personal profile, which you will write in step three. Your personal profile will emphasize key skill sets. These skill sets will be used as subheadings in the professional experience heading, which will be the section listed immediately after the personal profile. When writing a personal profile, include key general skills and qualities desired by your target employer. Related work experience, specific skills, important activities, and significant accomplishments will be detailed under each respective subheading. Share major accomplishments and responsibilities from each position listed in your work history. Education and then work history are listed after professional experience. If necessary, add a second page to your résumé. Refer to Figure 14.4 for the advanced skill set résumé layout and Figure 14.5 for an example of an advanced skill set résumé.

For both résumé layouts, present employment history and education in reverse chronological order (most recent job first). When listing work history, bold the job title, not the place of employment. Only the month and year should be used when listing dates of employment. Be consistent in how dates are listed on the résumé.

Once you have determined which résumé layout is best for your current situation, in your electronic document, arrange the information gathered in step one into the correct résumé layout. Avoid résumé templates; these can be difficult to update, modify, and personalize.

Step Three: Writing a Skills Summary or Personal Profile

A foundation for both your job search strategy and building a winning résumé is to write a skills summary statement or personal profile. A **skills summary** is an introductory written statement for individuals with little or no work experience and is used on a functional résumé. The summary statement replaces the traditional one-sentence career objective. A skills summary statement encapsulates a job seeker's knowledge and skills in a brief statement that communicates his or her career objective while highlighting the value an individual brings to a prospective employer. A **personal profile** is an introductory written statement for individuals with professional experience related to their target career and is used on an advanced skill set résumé. The skills summary or personal profile will be the first item listed on your résumé following the information heading.

Employers spend little time looking at the entire résumé, and most of that time is spent looking at the top of the résumé. A good summary or profile is a way to make your résumé personalized and powerful. The responses from your completed accomplishments worksheet and career assessment from Chapter 13 provide a focused summary of your current career goal based on the knowledge, skills, and abilities you possess for a specific job. Depending on the layout of your résumé, this information will either have the heading "Skills Summary" or "Personal Profile." Make your skills summary or personal profile specific to the employer and job for which you are applying to increase your chances of being considered for the job. The skills summary or personal profile is the only place on a résumé where it is acceptable to use the words "I" and "my."

Those with little or no work experience in a targeted career will utilize the skills summary. In creating a skills summary, include your target job, employer, and key skills and experience into a one- to two-sentence statement.

Skills Summary Examples

- **Skills Summary:** Highly motivated and positive person seeking to obtain a position as an Office Professional with Roxy's Clothing Company that will enable me to utilize my current customer service skills and office assistant education.
- **Skills Summary:** To obtain a Medical Assistant Clinician position at Healthcorp, where I can demonstrate and increase my current medical assisting skills such as pharmacology, laboratory, and diagnostic procedures.

Those with extensive work experience will utilize a personal profile. In creating a personal profile, review your key skills and accomplishments and group these items into general categories. Also identify key qualities you possess that are required for your target job. Take this information and turn it into a two- to three-sentence statement that provides a snapshot of your professional qualifications in a manner that sells your knowledge, skills, and abilities.

Personal Profile Example

- **Personal Profile:** Highly professional and detail-oriented accounting professional with demonstrated leadership and success in the areas of payroll, collections, and project management. Possess excellent analytical, communication, computer, and organizational skills. Bilingual English/Spanish (read, write, and speak).

Exercise 14.1

Write a draft of your skills summary or personal profile.

_____

_____

_____

Step Four: Insert Skills, Accomplishments, and Experience

After arranging your information into the correct layout and inserting your career summary or profile, it is time to detail the information listed in your skills, work experience, and professional accomplishments. List skills relevant to your target job first. Be specific when referring to common workplace skills. For example, computer skills are too general and typically include many different areas such as networking, programming, applications, data processing, and/or repair. An employer needs to know what specific computer skills you possess and the proficiency level (e.g. basic, intermediate, or advanced) with specific software. Work experience includes learned skills, job duties, and accomplishments from paid, unpaid, and volunteer work. When presenting work experience, list the job title, company name, city, state, and duties of the position. Make every effort to not state "responsible for . . ." Instead, list specific accomplishments. Accomplishments are activities you achieved beyond your job duties. Do not assume the reader will know what you have done. Whenever possible, quantify outcomes related to your skills, responsibilities, and professional accomplishments. For example, if your duties included working with others, phrase your duties to read "Worked with team of 12 to assist over 350 customers daily."

As you insert professional accomplishments and responsibilities into your electronic file, include job-specific, transferable, and soft skills. **Job-specific skills** are those that are directly related to a specific job or industry. If you were to change careers, job-specific skills would probably not be useful. For example, if you are a medical billing clerk who knows how to use a specific software program such as Medical Manager, you will not use this skill if you become an elementary school teacher.

Transferable skills are skills that are transferred from one job to the next. Should you change careers, you will still be able to use these skills. For example, you learned how to input data into a computer for billing purposes. If you then become an elementary school teacher, you will use these keyboarding skills when you report student data. Speaking a foreign language is an excellent transferable skill. If you are bilingual (speak or write a second language), include this information in your résumé. Inform the employer if you read, write, or only speak that second language.

The term **soft skills** refers to the people skills necessary when working with others. Employers want employees that are reliable, team players, good communicators, and able to get along well with others. Employers need employees with job-specific, transferable, and soft skills; therefore, list all these skills on your résumé.

Résumés do not normally contain complete sentences. They contain statements that sell your skills, qualifications, and work experience. Except for the skills summary or personal profile, the words "I" and "my" should not appear.

Exercise 14.2

List as many job-specific, transferable, and soft skills as possible. If you have not yet held a job, list job skills you have developed from academic studies or volunteer work.

_____

_____

_____

Organize your skills and work experience by first listing the key skills required for your target job. When communicating your skills, experiences, and accomplishments, write with energy. Use action verbs. Action verbs are also referred to as power words and are used to describe your accomplishments in a lively and specific way. For example:

Instead of: "I started a new accounts receivable system."

Write: "Developed a new accounts receivable system that reduced turnaround time by 20 percent."

Or

Instead of: "Updated electronic medical records."

Write: "Converted paper to electronic medical files for a five-member physician office."

Refer to Table 13.1 for a list of sample power words and Table 14.1 for sample power statements. When using power statements, quantify your accomplishments whenever possible.

Exercise 14.3

Refer to the accomplishments worksheet completed in Activity 13.1. Review these accomplishments and turn them into powerful statements. Quantify whenever possible.

_____

_____

_____

Table 14.1 Power Statements

Assisted in preparing, cooking, garnishing, and presenting food for _____

Inspected quality of incoming supplies and maintained inventory for _____

Established and maintained positive and effective working relationships with _____

Ordered, maintained, and trained in the use of _____

Knowledgeable of _____ software

Recorded patient history and vitals for _____

Cleaned and sterilized instruments and disposed of contaminated supplies for _____

Built website for _____

Developed technology protocols for _____

Wrote copy that was utilized in _____

Designed new brand identity for _____

Evaluated code to update existing site for _____

Analyzed, planned, and repaired _____

Prepared and processed _____

Organized and maintained records for _____

Drafted and typed business correspondence for _____

Scheduled and distributed _____

Step Five: Review the Completed Résumé

Prior to finalizing your résumé, ensure you have added all information identified in steps one through four to your electronic document. As you finalize your résumé, check for information that is frequently forgotten or not presented appropriately. Confirm your information heading contains complete contact information.

Carefully evaluate the skills summary or personal profile to ensure it introduces the reader to who you are and motivates him or her to learn more about your specific knowledge, skills, abilities, and key accomplishments.

In step two, you determined whether a functional or advanced skills set résumé layout was appropriate for your situation. Review the respective layout for proper heading order and refer to the sample résumés. Confirm that your experience and education are listed in reverse chronological order (most recent first). Keep your résumé consistent in its presentation, including all periods or no periods at the end of each line, line spacing, alignment of dates, date format, bold/italics, upper- and lowercase words, underlines, and other formatting. Also, review for consistency with the use of tense in each section (e.g., -ing and -ed) and with the use of the postal abbreviation for your state (e.g., the state is CA, not Ca., not Ca, not C.A.). When your draft résumé is complete, spell-check and proofread the document to ensure it is free of typographical errors and inconsistencies.

Underlines, bold, and italic print are acceptable for emphasis but should not be overdone. Avoid using bullets throughout your résumé; use bullets only to emphasize key areas. Only use small round or square bullets. Fonts and sizes should be easy to read. Times New Roman or Arial are most common. Apart from your name on the information heading, do not use more than two different font sizes (no smaller than 11 point and no larger than 14 point). Unless you are in the graphic design industry, avoid using different color fonts, highlights, or graphics on your résumé; use only black ink. It is not appropriate to state, "References Available Upon Request" at the close of your résumé. Professional references are to be listed on a separate sheet and provided only when requested. Refer to Chapter 13 for proper format for a professional reference list.

Once you have made sure your résumé is presented professionally, is free of errors, and does not contain unnecessary or inappropriate information, print the résumé in black ink on 8½ × 11–inch, letter-sized paper. Double-sided résumés are not appropriate. If your résumé is more than one page, place your name at the top of each page after page one. Proper résumé paper is cotton-fiber, 24-pound white or off-white (not bond) paper of good quality. Colored paper, especially if dark, is both difficult to read and does not photocopy well. Do not use fancy paper stocks or binders. Also, avoid stapling your résumé or other job search documents. Since résumés are frequently photocopied, stapled résumés and other job search documents may be torn in the process.

When you have completed your résumé and believe it is ready for distribution, have several individuals whom you trust review it for clarity, consistency, punctuation, grammar, typographical errors, and other potential mistakes. Remember that complete sentences are not necessary and, with the exception of your career objective, the words "I" or "my" should not be used. Your résumé must create a positive, professional visual image and be easy to read.

Sharing Your Résumé

As you prepare to share your completed résumé with both potential employers and members of your professional network, you may have the option of presenting your résumé on résumé paper (traditional hard copy) or electronically (online). Your traditional résumé contains key information necessary for sharing with all potential employers. When sharing your résumé online, ensure it contains key words relevant to your target job. Employers and job boards commonly drop résumés into a database or résumé tracking system that allows recruiters to search for potential applicants based on key words and phrases that match the position they are trying to fill. When posting an online résumé, you may be required to cut and paste sections from your traditionally formatted résumé. During this process, you may lose the formatting. Do not worry. Visual appeal is not an issue in this process. You are merely dropping your information into a database. Your focus should be on utilizing key words and phrases related to the target job that concisely sell your skills and quantify your accomplishments.

Exercise 14.4

Circle the inconsistencies and errors on the following résumé.

1100 EAST FAVOR AVENUE • POSTVILLE, PA 16722
PHONE (555) 698-2222 - E-MAIL AERIE@PBCC.COM

AMANDA J. ERIE

OBJECTIVE

Seeking a position as an Administrative Assistant where I can utilize my office skills to better my career

SUMMARY OF QUALIFICATIONS

- Computer software skills include Microsoft Word, Excel, Outlook, Access, and Power point
- Knowledge of Multi-line telephone system, filing, data entry, formatting of documents and reports, and operation of office equipment.
- Excellent interpersonal skills and polished office etiquette.
- written and oral communication skills
- Typing skills
- Bilingual

EDUCATION

Reese Community College, Postville, PA Currently pursuing AA Degree in Office Occupations.

Calvin Institute of Technology, Cambridge, OH Office Technology Certificate Spring 2010

WORK AND VOLUNTEER EXPERIENCE

01/11 – Present   *Rigal Entertainment Group*   Postville, CA
Usher – Responsible for ensuring payment of services. Answer customer inquiries. Collect and count ticket stubs.

11/07 – 02/09   Lablaws   Cambridge, OH
Cashier – Operated cash register, stocking, assisting customers

Jan/07 – 04/07   Jolene's Diner   Cambridge, OH
Server – Provided customer service by waiting tables, cleaned, and operated cash register

The second consideration when converting a traditional résumé to an online version is sending it as an attachment to preserve formatting. If you are sending your résumé as an attachment, it is best to send it as a portable document file (.pdf). Doing so ensures that the résumé layout is properly maintained through the file transfer, and also ensures that those who do not use the same word processing software are able to read the file.

Most colleges and career centers host electronic job boards that allow students to upload their résumés for recruiters and employers who are seeking to hire students. Make use of this valuable resource. There are also many job boards specific to industries. Another popular means of sharing an electronic résumé is through social media sites. Make certain you are posting your information on valid business and industry sites and not personal sites. As with a traditional job search, track and monitor all activity related to your online search.

As stated earlier, résumés that are posted online are frequently dropped into databases and scanned for key words that match qualifications/skills required for current job openings. Therefore, ensure you thoroughly research and include key words on your résumé for targeted positions. Doing so will increase the probability that your résumé will match the target job posting.

When posting a résumé online, date your résumé and update it every two to three months. Most employers won't view online résumés that are more than six months old. Just as with a hard-copy résumé, protect your identity and do not share personal information of any kind, including marital status, birthdates, or your Social Security Number.

Cover Letters

A **cover letter** serves as an introduction to your résumé and is often the first impression a potential employer will have of you. Employers frequently use cover letters as screening tools. Even when limiting your job search to online venues, create a cover letter and use the body of the letter as the primary content of your e-mail message. Whenever possible, tailor your cover letter by including information found when researching the company and position for which you are applying.

When writing a cover letter, convey a friendly yet professional tone using complete sentences and proper grammar. The goal of a cover letter is to communicate how your key skills, experience, and accomplishments can meet the employer's needs. A basic cover letter contains three paragraphs. The first paragraph contains the purpose of the letter, the specific position for which you are applying, and how you learned of the position. If you have a contact within your target company, share the name of this individual and refer to how that individual informed you of the open position. In one sentence, summarize why you are interested and/or qualified for the position. Finally, share why you are interested in the organization, indicating any research you have conducted on the position and/or employer.

The second paragraph refers to the attached résumé and highlights the skills and qualifications you possess that the employer is requesting for the target job. Summarize how your key skills and qualifications match the employer's needs. Communicate what you can offer the company, not what you want from the company. Do not duplicate what is already listed on your résumé; instead, emphasize your experience and key skills.

Although it is acceptable to use the words "I" and "my" in a cover letter, be cautious to not begin most of your sentences with the word "I." Instead, focus the attention toward the employer by placing the company first and making its needs the priority. For example:

> Instead of writing, "I am proficient in the most recent version of Word." Write, "Your company will benefit from my proficiency in the most recent version of Word."

The purpose of the final paragraph is to request an interview (not the job). Do not state that you look forward to the employer contacting you; instead display initiative by stating that you will follow up on your request for an interview within a week. Include your phone number and e-mail contact information, even though it is already included in your information heading. Close courteously and include an enclosure notation for your résumé.

Do not address your cover letter to a department, the company name, or "to whom it may concern." Address the cover letter to a specific person, ideally to the person who will be making the hiring decision. This is typically the individual who directly supervises the target position. Research or call the company to identify a specific name and title, including the appropriate spelling and gender. If you still cannot secure a specific name, use a subject line instead of a salutation. Instead of "To Whom It May Concern," write "Subject: Account Clerk Position." Use the proper business letter format for your cover letter presented in Chapter 9. Each word and paragraph in your cover letter must have a purpose. Your goal is to communicate how your knowledge, skills, abilities, and accomplishments fill a targeted company's needs and make the reader want to review your résumé. The cover letter setup and sample cover letter in Figure 14.6 will assist you in creating a winning cover letter.

Cover Letter Setup and Example, see Figure 14.6 on pages 224

Print the cover letter on the same paper used for your résumé. Use the same information heading you created for your résumé to create a consistent and professional visual appeal for your résumé package. Avoid common mistakes such as typographical or grammatical errors, forgetting to include a date, or forgetting to sign the cover letter. When sharing your résumé package via the Internet, utilize an electronic signature on your cover letter. As with your résumé, have someone you trust proofread your letter before you send it to a potential employer. Any error communicates a lack of attention to detail and has the potential to disqualify you from securing an interview.

To complete your résumé portfolio, use the same information heading from your résumé and cover letter and add it to the reference list you created in Chapter 13. Refer to Figure 14.7 for an example of a professional reference list.

Reference Page Example, see Figure 14.7 on page 225

Tailoring Your Résumé and Cover Letter

As previously stated, tailor both your résumé and cover letter specifically to each job and company for which you are applying. Carefully review the target job announcement and identify key job skills that the position requires. If possible, secure a copy of the job description from the company's human resource department. In addition, conduct an occupational quick search on the ONET database. This database was developed for the U.S. Department of Labor and provides key information by job title. Add the key knowledge, skills, and abilities from your ONET research to those the employer desires and then emphasize this information on your résumé. If necessary, rearrange the order of the information presented on your résumé so the key skills required for your target position are presented first. Also, emphasize your specific qualifications that match those required for the open position on your cover letter.

List a daytime phone number and e-mail address on both the cover letter and résumé. Because most invitations for job interviews occur over the phone, your voice mail message needs to be professional. Do not include musical introductions or any other greeting that does not make a positive first impression to a potential employer. As mentioned in Chapter 13, maintain a professional e-mail address to use in your job search.

Topic Situation

Emily's friend Rebecca was a practical joker. Emily enjoyed calling Rebecca because her voice mail message started with a joke or had some strange voice and/or music. However, the last time she called Rebecca, she noticed that her message was normal. The next time Emily saw Rebecca, she asked her why her voice message was suddenly so serious. Rebecca explained that she had recently applied for a job and had been selected to interview. However, she was embarrassed because when the interviewer called to arrange the appointment, he left a message suggesting that Rebecca change her voice mail to be more professional.

TOPIC RESPONSE

What would be an appropriate voice mail message for Rebecca? Should Rebecca acknowledge the interviewer's voice mail recommendation when she goes into the interview?

Tips for Special Circumstances

Gaps in work experience on your résumé will sometimes be a red flag for an employer. It is common for an individual to have time gaps in a résumé as a result of staying at home to raise a young child, care for an elderly relative, or continue his or her education. Many have time gaps simply from seeking employment. If the time gap is less than a year, list only years instead of the months and years. If the gap is longer than a year, individuals should use the advanced skill set résumé format. Identify a key skill you sharpened during the time gap and relate this experience to a key skill necessary for your target job and industry. For example, if you stayed at home to care for an elderly relative, the experience most likely improved your time management and organizational skills. You also most likely improved your awareness of diverse populations, including the elderly and disabled. Without revealing personal information, list your activity as you would any other job on your résumé. You need to be prepared to answer questions about gaps in the interview. Strategies for answering questions such as these will be covered in Chapter 15.

If you are an ex-offender and have served time in prison and are now attempting to reenter the workforce, you are to be congratulated for wanting to move forward with your life. By succeeding in college you have shown you made restitution for your past choices. Be honest with the potential employer.

On your résumé, include all jobs you have held and skills you learned while incarcerated. For the jobs you held while incarcerated, list the correctional facility in place of the employer. Also, list all education, including degrees and courses you received while making restitution along with the educational institution that provided the courses.

The employment application is a legal document. At the bottom of this document, applicants sign a statement affirming that all information provided on the application is true. Therefore, you must not lie. If, after being hired, your employer discovers that you have lied on the application, you may face immediate termination. The majority of applications ask if you have been convicted of a felony. Please note that arrests are not convictions. If you have been convicted of a felony, check "Yes." The application should also provide a space to write a statement after the felony question. Do not leave this space blank. In this space, write, "Will explain in detail during interview."

MyStudentSuccessLab Please visit **MyStudentSuccessLab**: Anderson|Bolt, Professionalism Skills for Workplace Success, 4/e for additional activities, resources, and outcomes assessments.

Workplace Dos and Don'ts

Do keep your résumé updated with skills and accomplishments	*Don't* wait until the last minute to update your résumé
Do change your résumé layout after you have advanced skill sets	*Don't* lie on your résumé or cover letter
Do pay attention to formatting details on your résumé and cover letter	*Don't* distribute a résumé or cover letter that has not been proofread by someone you can trust
Do ensure your résumé and cover letter are free of errors prior to sharing them to employers	*Don't* forget to sign your cover letter

YOUR NAME (16 point, bold)
Your Address (12 or 14 point, bold) ■ City, State, Zip
■ **Phone Number** (Give only one number and include area code)
Email Address (Remove hyperlink)

Section headings can be left or centered, 12- or 14-point font, and uppercase or initial cap. Format headings consistently throughout the résumé. Keep spacing equal between each section.

Horizontal line optional, weight varies

SKILLS SUMMARY

Headings can be on the left or centered, 12- or 14-point font, and uppercase or initial cap, but should be formatted consistently throughout the résumé. Keep spacing equal between each section

QUALIFICATIONS (OR SKILLS)

- Relate to target job, use job-related skills and transferable skills
- Most relative to the job are listed first
- Bullet (small round or small square only) these items to stand out

Emphasize skills and education before work experience.

EDUCATION

You may list before qualifications

List schools in chronological order, most recent attended first

Do not list high school if you graduated from college (if listed, do not include dates)

Include the years attended college or pending graduation date

WORK EXPERIENCE

Name of Company and City, State—No Addresses

Job title, dates employed (month, year)

List the jobs in chronological order with most recent first

List the duties, responsibilities, and achievements and be consistent in your setup by using same tense throughout (-*ed* or -*ing*)

OTHER CAPABILITIES

Optional items in this section such as honors or awards may not be directly related to the job but may interest the employer.

- Watch periods, punctuation, and spelling
- Align all dates and format them consistently
- Keep to one page
- Use a regular 12-point font (except heading), no color
- Use résumé paper, no dark or bright colors
- Do not use full sentences or *I, me,* or *my*
- References are not necessary; you will list these on a separate sheet

Figure 14.1

Functional Résumé Layout

Cory S. Kringle
4 Success Lane, Fresno, CA 93702
555-123-4567
ckringle15@careersuccess.job

SKILLS SUMMARY

Highly motivated and positive person seeking to obtain a position as an Office Professional with Roxy's Clothing Company that will enable me to utilize my current customer service skills and office assistant education.

QUALIFICATIONS

- Type 50 wpm with 97% accuracy
- Intermediate skills using Microsoft Word, Excel, Access, PowerPoint, and Outlook
- Accurately proofread and edit documents
- Knowledge in records management; traditional and computerized
- Positive telephone skills
- Excellent oral and written communications skills
- Positive attitude, motivated, and organized
- Excellent and positive customer service skills

EDUCATION

2013–2015	Fresno Community College	Fresno, CA

Associate of Art Degree, Business & Technology
Office Professional Certificate
GPA 3.9, Dean's list

Courses

- Working Relations
- Business English
- Records Management (traditional and computerized)
- Today's Office
- Document Processing using Microsoft Word
- Computer Applications using Microsoft Office

EXPERIENCE

06/2013–present	Fine Linens by Jen	Fresno, CA

Cashier

Provide customer service including cashiering and returns in a busy setting, serving an average of 200 customers a day. Assist team of 7 with inventory, processing and placing merchandise on the floor, helping return go backs, stocking of merchandise in back/ stockroom, and training new hires.

Figure 14.2

Functional Résumé
Example without Career
Work Experience

CORY S. KRINGLE

4 Success Lane ▼ Fresno, CA 93702 ▲ 555-123-4567 ▼
ckringle15@careersuccess.job

SKILLS SUMMARY

Newly licensed and motivated utility technician professional seeking an entry-level gas utility position. With training in mechanical aptitude and schematic reading I am ready to be a part of the success at Great Utility Company.

KEY SKILLS AND QUALIFICATIONS

- Computer literacy and applications including MS Word and Excel
- Knowledge and practice in mapping and schematic reading
- Certification in gas theory, principles, and technical skills
- Practice using occupational safety, health, and environmental concerns
- Introduction to electronics and industrial mechanics
- Highly ethical
- Responsible
- Detail oriented
- Excellent communication skills
- Knowledgeable of governmental regulations and safety standards

EDUCATION AND CERTIFICATIONS

Yu Technical College, Meadville, CA   12/15
Associate of Science Degree—Utility Technician

WORK HISTORY

Gas Apprentice—residential wiring   8/15–12/15
Pennsylvania Local #7777, Meadville, CA
Under the direction of a journeyman, worked indoors and outdoors with gas lines. Checked for gas leaks, repaired gas lines, and installed and connected gas lines for both residential and industrial sites. Read meters, installed meters, and adjusted meters. Tested electrical systems and continuity of wiring, equipment, and fixtures utilizing current technologies.

Grocery Clerk/Assistant Butcher
Great Foods, Meadville, CA   5/13–8/15
Assisted three butchers in meat department weighing, pricing, and wrapping products. Cut and prepared product for cases. Applied excellent customer service in taking and filling orders. Served as cashier and front duties when necessary. Cleaning and secured area for store closing.

Figure 14.3

Functional Résumé
Example with Career
Work Experience

YOUR NAME (16 point, bold—include middle initial)

Address (12 or 14 point, bold) ■ **City, State Zip** ■ **Phone Number** (Give only one number and include area code)

E-Mail Address (Remove hyperlink)

PERSONAL PROFILE:

> Section headings can be left or centered, 12- or 14-point font, and uppercase or initial cap. Format headings consistent throughout the résumé. Keep spacing equal between each section.

In a three- to four-sentence statement, insert key skills and accomplishments related to the target job. Group key skill sets into categories that will be used as skill set subheadings in the Professional Experience section. Include key qualities you possess required for the target job.

PROFESSIONAL EXPERIENCE:

> Use the skill set categories listed in the Personal Profile as subheadings. Under each skill set subheading, elaborate on key skills, experience, and accomplishments related to each category. On an advanced skill set résumé, Professional Experience is listed before Education and Employment History.

First Skill Set Subheading

- Communicate experience and key accomplishments relating to your first skill set subheading
- Using power words, quantify as much as possible
- Include duties, responsibilities, and achievements

Second Skill Set Subheading

- Communicate experience and key accomplishments related to your second skill set subheading
- Accomplishments and experience most relative to target job are listed first
- Communicate both job-related and transferable skills

Third Skill Set Subheading

- Communicate experience and key accomplishments related to your third skill set subheading
- Use consistent tense for statements (-ing for present tense, -ed for past tense)
- Do not use *I, me,* or *my*
- Complete sentences are not necessary

WORK HISTORY:

Name of Company, City, State (do not include address)   Dates Employed (month & year)
Job Title (bold the job title, NOT the employer)
List jobs in reverse chronological order

> Proofread for proper punctuation and spelling. Advanced skill set résumés can be one or two pages. If more than one page, place name at the top of second page. Avoid color and graphics. Utilize consistent alignment of bullets and dates.

EDUCATION AND LICENSES:

Degree
College, City, State (do not include address)   Date Degree Awarded (month & year)
License, State, or Organization Awarding Certification   Date of Award (month & year)
Degree College, City, State (do not include address)   Date Degree Awarded (month & year)

PROFESSIONAL AFFILIATIONS/CERTIFICATIONS:

Name of Organization, status (member, officer, etc.)   Dates of Membership
Certification, Certifying Organization   Date of Certification
Volunteer or Service Activities   Dates of Service (month & year)

Figure 14.4

Advanced Skill Set
Résumé Layout

CORY S. KRINGLE

4 Success Lane ■ Fresno, CA 93702 ■ 555-123-4567 ■ ckringle15@careersuccess.job

PERSONAL PROFILE

Results- and efficiency-focused professional with experience in sales/vendor relations, inventory/warehousing, and management/supervision. Proven ability in relationship management with demonstrated and consistent increase in sales over a five-year period. Inventory expertise includes streamlined operations, improved productivity, and favorable inventory ratio utilization for wholesale food supplier. Management ability to create goal-driven teams, groom leaders, and facilitate the creation of a learning organization.

PROFESSIONAL EXPERIENCE

Customer Service Orientation ■ Innovative Risk Taker ■ Excellent Quantitative Skills Purchasing, Inventory Planning, & Control ■ Supply Chain Management ■ Warehouse Operations ■ Process Improvement ■ Cost Containment ■ Hiring, Staffing & Scheduling Safety Training ■ Excellent Computer Knowledge

Sales/Vendor Relations

- Through the establishment of vendor relationships, schedule product installations, exchanges, buy-backs or removals of equipment or other assets including supplier networks and agent contacts in order to meet customer expectations for private soda company, have grown sales territory from a two-county area to (highly profitable) tri-state contract area over four-year period.
- Source and facilitate delivery of product (e.g., beverage equipment, parts, point of sale material, return of assets) for retail suppliers. Sales complaints are consistently under 5% per year, while sales volume and customer satisfaction rates are the highest of all national sales teams and consistently continue to grow.
- Research and resolve issues for customers, business partners, and company associates in order to expedite service, installations, and/or orders utilizing information systems and working with 12 regional supply chain partners.
- Create and maintain partnerships with customers, clients, or third-party service providers (e.g., contract service/installation agents, distributors) by establishing common goals, objectives, and performance target requirements in order to improve customer service and satisfaction (which is currently 98.7%) for clients that are my direct responsibility.
- Developed troubleshooting equipment process which allows retail suppliers to receive immediate response on service issues (e.g., beverage vending, dispensing) via telephone or Internet to minimize customer down time and service cost.

Figure 14.5

Advanced Skill Set
Résumé Example

Cory S. Kringle

Page Two

Inventory/Warehousing

- Maintain customer contact to confirm service or orders and to ensure accuracy with equipment repairs, product deliveries, and routine service scheduling for regional food service broker who was responsible for 65% of annual company revenue.
- Receive, record, and respond to customer inquiries using specially designed database which documents best practices from nationwide food service association to provide improved service, order accuracy, and optimized supply chain efficiency.
- Process daily orders for goods and services with over 20 food service business partners, customers, suppliers, and company associates, either through direct telephone contact or electronic means, to increase speed and accuracy of order transactions and improve loss prevention systems.

Management/Supervision

- Developed and trained team of 20 on inventory control, customer service, and safety for local food service provider. Program was so successful customers within the company supply chain requested and received training. To date, over 500 individuals have received custom training.
- Supervised cross-functional team of 100 including order technicians, outside repair personnel, transportation associates, warehouse attendants, and loss prevention specialists.
- As assistant manager for college-town restaurant, assisted in the hiring, training, scheduling, and performance evaluation of staff for small soda company and local food service supplier.

WORK HISTORY

Connor Cola Company, Susanville, CA **Vendor Relations Associate**	2012–present
Elizabeth Food Service, Pocatoe, NE **Warehouse Manager**	2009–2012
Mango and Carolyn Ribs'N Stuff, Pocatoe, NE **Assistant Restaurant Manager**	2006–2009

EDUCATION/PROFESSIONAL DEVELOPMENT

University of Nebraska, Lincoln, NE Bachelor of Science, Business Management/Marketing	2012

Figure 14.5

(*continued*)

Utilize the same heading as your résumé to create a consistent, professional presentation.

Utilize the business letter format presented in Chapter 9.

Tailor cover letter and address to a specific individual. Ensure correct spelling and title.

Personalize the salutation.

First paragraph includes the purpose of the letter, position applying for, and how you learned of the position. Summarize why you are interested and qualified, and indicate any research conducted.

Second paragraph refers to résumé and highlights skills and qualifications you possess for the target job. Summarize how your key skills and qualifications match the employer's needs. Limit the use of *I* and *my.* Communicate what you can offer the company, not what you want from the company. Do not duplicate what is already listed on the résumé; emphasize experience and key skills.

The closing paragraph requests an interview. Display initiative by stating that you will follow up on a request to interview within a week. Include your phone number and e-mail.

Use a courteous closing. Sign your name (if appropriate, use an electronic signature). Utilize proper business-letter formatting.

CORY S. KRINGLE
4 Success Lane, Fresno, CA 93702
555-123-4567 ckringle15@careersuccess.job

September 22, 2015

Anita Stephens, HR Manager
Clay Office Supplies
435 East Chesny Street
Fresno, CA 91188

Dear Ms. Stephens:

I recently spoke with Terry Moody, a production manager for Clay Office Supplies, and he recommended that I forward you a copy of my résumé. Knowing the requirements for the position and that I recently received my degree in office occupations, Mr. Moody felt that I would be an ideal candidate for your Office Assistant position.

I would welcome the opportunity to be employed at Clay's Office Supplies since this is the largest and best-known office supply company in the city. Your company has a reputation of providing excellent products and service, which is why Clay's Office Supplies would benefit from my knowledge and skills. I am accustomed to and thrive in a fast-paced environment where deadlines are a priority and handling multiple jobs simultaneously is the norm. As you can see on the attached résumé, my previous job required me to be well organized, accurate, and friendly. My educational courses taught me how to utilize current skills and technologies and sharpened my attention to detail. I enjoy a challenge and want to contribute to the success of your company.

Nothing would please me more than to be a part of your team. I would like very much to discuss with you how I can benefit Clay Office Supplies. I will contact you next week to arrange an interview. In the interim, I can be reached at 555-123-4567 or at ckringle15@careersuccess.job.

Sincerely,
Cory S. Kringle

Cory S. Kringle
Enclosure

Figure 14.6

Cover Letter Setup and Example

CORY S. KRINGLE
4 Success Lane, Fresno, CA 93702
555-123-4567 ckringle15@careersuccess.job

PROFESSIONAL REFERENCE LIST

Name	Relationship	Phone	E-mail	Mailing Address
Autumn Hart	Former Instructor, Yu Technical College	555.555-1111	atmnhrt@yutc.scl	123 Hillvalley Meadville, CA
Gloria Montes	Owner, Fine Linens by Jen	555.555-1112	gloria@linens.sleep	5432 Food Ct. Fresno, CA
Gary Solis	Manager, Conner Cola	555.555-1113	solisg@conner.cola	2220 Tulare Susanville, CA
Patty Negoro	Owner, Mango Ribs	444.555-1114	pattyn@eatribs.com	444 Adoline Pocatoe, NE

Figure 14.7

Reference Page Example

Concept Review and Application

You are a Successful Student if you:

- Write a career objective/personal profile
- Write a résumé
- Write a cover letter

Summary of Key Concepts

- A winning résumé makes it easy for potential employers to quickly and easily identify your skills and experience.
- Update your résumé with new skills and accomplishments at least once a year.

- Include both job-specific skills and transferable skills on your résumé.
- Use the correct résumé layout for your career work experience.
- A cover letter is most often an employer's first impression of you.
- Check that your résumé and cover letter are free of typographical and grammatical errors.
- Share your résumé electronically as a .pdf file to ensure the résumé layout is maintained.

Self-Quiz MATCHING KEY TERMS

Match the key term to the definition using the identifying number.

Key Terms	Answer	Definitions
Advanced skills set résumé layout		1. An introductory written statement used on a functional résumé for individuals with little or no work experience
Cover letter		2. An introductory written statement used on an advanced skill set résumé for individuals with professional experience related to their target career
Functional résumé layout		3. A formal written profile that presents a person's knowledge, skills, and abilities to potential employers
Information heading		4. A letter that introduces your résumé
Job-specific skills		5. Skills that can be transferred from one job to another
Personal profile		6. A résumé layout that emphasizes relevant skills when related work experience is lacking
Résumé		7. People skills that are necessary when working with others
Skills summary		8. Skills that are directly related to a specific job or industry
Soft skills		9. A résumé heading that contains relevant contact information including name, mailing address, city, state, ZIP code, contact phone, and e-mail address
Transferable skills		10. A résumé layout used by those with extensive career experience that emphasizes related work experience, skills, and significant accomplishments

Think Like a Boss

1. What would you look for first when reviewing a résumé?
2. What would your reaction be if you were reading a cover letter that had several spelling and grammar errors?

Activities

Activity 14.1

Complete the following table.

Education (list most recent first)				
School Name	**City, State**	**Dates**	**Degree, Certificate, Credential, License**	**GPA**

Skills

Employment (list most recent first)			
Employer	**Employment Dates**	**Job Title**	**Duties**

Languages	Fluency (Read, Write, and/or Speak)

Honors and Awards	Dates	Place

Professional/Community Involvement		

Activity 14.2

Conduct an Internet search to identify résumé power words/phrases. List at least five new words that are not in the text.

1. _____

2. _____

3. _____

4. _____

5. _____

Activity 14.3

Search for a specific job you would like to have when you graduate and then complete the following table. This information will be used to tailor your résumé and cover letter.

Position for which you are applying	
How you learned about the job	
Any contact you have had with the employer or others about the job	
Why are you interested in this job?	
Why are you interested in this company?	
What products or services are provided?	
List relevant skills related to the job description	
List reasons this company should hire you	
Indicate your desire for an interview	
Indicate your flexibility for an interview (time and place)	

Activity 14.4

Using a word processing program and the steps and/or exercises from this chapter, create a résumé for the job you found in Activity 14.3.

Activity 14.5

Using a word processing program and the information from this chapter, create a cover letter for the job you found in Activity 14.3.

15

Interview Techniques

prepared • poised • present

After studying these topics, you will benefit by:

- Implementing pre-interview strategies and activities
- Conducting company- and job-specific research for interview preparation
- Creating a powerful and unique personal commercial
- Compiling an interview portfolio and e-portfolio
- Practicing interview techniques and appropriate responses to common interview questions
- Implementing pre-interview preparation activities
- Demonstrating winning behavior during face-to-face and technology-based interviews
- Naming and describing common interview methods and types of interview questions
- Explaining key areas of employee rights and knowing how to respond to discriminatory questions
- Formulating appropriate responses to special circumstances and tough questions
- Preparing for post-interview activities including salary negotiation, employment screenings, tests, and medical exams

HOW DO YOU RATE?

Do you know proper interview techniques?	True	False
1. It is unprofessional to arrive more than 10 minutes early to the office where your interview is to take place.	☐	☐
2. It is best to have a draft of a post-interview thank-you note written prior to an interview.	☐	☐
3. The same amount of pre-interview preparation should be made for an Internet and/or telephone interview as is made for a traditional face-to-face interview.	☐	☐
4. Employers expect a job candidate to ask relevant questions during interviews.	☐	☐
5. When offered a job, it is acceptable to negotiate a salary.	☐	☐

▶ If you answered "true" to the majority of these questions, congratulations. You are already aware of effective interview techniques and are ready to successfully interview.

The Interview

You've conducted a targeted job search and created and distributed your résumé, and now it is time to interview. A successful interview involves more than dressing sharp. It includes advance preparation, confidence, and a strategy to be used before, during, and after this important meeting. During an interview, an employer is looking to hire the best person to represent his or her company. Your goal is to communicate visually and verbally that you are the right person for the job. A job search takes work, takes time, and can sometimes be frustrating. Do not be discouraged if you do not get an interview or job offer on your first try. The purpose of this chapter is to provide you the skills and confidence to secure a good job in a reasonable time.

The Invitation to Interview

There is a strategy to successful interviews, and it starts as soon as you receive an invitation to interview. Most interview invitations are extended by phone or e-mail. Therefore, regularly check and immediately respond to both phone and electronic messages. This is a good reminder to maintain a professional voice mail message and e-mail address. When you are invited to interview, attempt to identify with whom you will be interviewing. You may be meeting with one person or a group of individuals. Your first interview may be a pre-screening interview where a human resource representative or some other representative from the company briefly meets with you to ensure you are qualified and are the right fit for the job.

During the invitation to interview, ask how much time the company has scheduled for the interview. If possible, also identify how many applicants are being asked to interview. If you are friendly, respectful, and professional, most companies will share this information. Attempt to arrange your interview at a time that puts you at an advantage over the other candidates. The first and last interviews are the most memorable, so try to secure one of these slots. If you are given a choice of times to interview, schedule your interview in the morning. People are much more alert at that time, and you will have a greater advantage of making a favorable and memorable impression. If this is not possible, try to be the last person interviewed prior to the lunch break or the first person interviewed immediately after the lunch break. Be aware that sometimes you will have no say in when your interview is scheduled so do not make demands. Politely ask the scheduler if it is possible for him or her to tell you who will be conducting the interview. Finally, note the name of the individual who is assisting you in arranging the interview. This will allow you to contact him or her should you need additional information and will also allow you to personally thank him or her should you meet on the day of the interview. The goal is to secure as much information as possible prior to the interview so you are prepared.

Exercise 15.1

Role-play an invitation to interview with another student.

_____

_____

_____

Company-Specific Research

There is no question that pre-interview research will assist you during a job interview. Unfortunately, many candidates ignore this step, thinking it is unnecessary or takes too much time. Prior to an interview, conduct research and identify as much as you can about the organization and the department of your target job. Not only will you have an advantage in the interview, but you will know if the company is the right fit for you and your career goals. Learn as much as you can about the company's leadership team, strategy, and any current event that may have affected the company. Review the company web and social network sites if available, or conduct a general Internet search to read blogs and other posts related to the company. Note products the company produces, identify the company's key competitors, and note any recent community activities or recognized accomplishments the company has been involved with.

In addition to the Internet, other sources for securing company information include company-produced brochures and literature, industry journals, and interviews with current employees and business leaders. Job-specific information is easily gathered by conducting a quick search on the ONET database using the position title as your keyword. As mentioned in Chapter 14, this database of occupational information provides key information by job title.

Use your research during the interview by mentioning specific information about the company. For example, a popular interview question is, "Why do you want to work for this company?" Be specific in your answer and respond with information that reflects your research. For example, say, "Your company has been green-conscious in the last two years, which is an area I believe is important," instead of saying, "I have heard it is a great company."

Topic Situation

Anita's friend Tomasz was excited about an interview he would be having in a week. When Tomasz was sharing his excitement with Anita, she asked him if he had conducted research on the company. Tomasz said he really didn't need to conduct research because the company was pretty well known. Anita explained that it is important to conduct research beyond general knowledge to make sure Tomasz stood out from the other candidates. They conducted an extensive Internet search on the target company and discovered useful information that Tomasz was able to use throughout his interview. After a successful interview, Tomasz thanked Anita and told her that the research prior to his interview gave him a lot of confidence that ultimately helped him secure the job.

TOPIC RESPONSE
What specific details should you identify when researching a company?

The Personal Commercial

Prepare a personal commercial that sells your skills and ties these skills to the specific job for which you are interviewing. A **personal commercial** is a brief career biography that conveys your career choice, knowledge, skills, strengths, abilities, and experiences that make you uniquely qualified for the position for which you are applying and your interest in the targeted position. Use the personal commercial at the beginning or end of an interview. Your goal is to sell yourself by communicating how your skills support the company and target job throughout the interview. Your personal commercial is essentially your "sales pitch" that communicates how you are the right candidate for the job.

Exercise 15.2

If you were alone in an elevator with the hiring manager of your target job, what key pieces of information about yourself would you communicate as you rode from the fifth floor to the first floor?

_____

_____

_____

Think About It

What hobby or unique skill unrelated to your target job is appropriate to use in your personal commercial?

In Chapter 13 you completed an accomplishments worksheet that assisted you in identifying your personal qualifications for a target job. This information was used to create a career objective for the résumé you built in Chapter 14. Use information from your résumé that highlights your personal accomplishments in your personal commercial. When writing your personal commercial, make it reflect your personality. Include your interest in your chosen career, activities related to the career, the skills you have acquired, and why you have enjoyed learning these skills. Also, when possible, share something unique about yourself that makes you stand out from other candidates and serves as a conversation starter for the employer.

In this one instance, a hobby or unique skill unrelated to your target job can be used. Do not include marital status or other potential discriminatory information. Your personal commercial should take no more than one minute to deliver. The following are examples of a personal commercial:

Medical Assistant Clinician	I just received my Medical Assistant Clinician Certificate from Success College. I enjoyed my courses and would like to ultimately become a nurse. As part of my training, I performed my required work experience at Community Hospital, where I was able to polish my clinical, medical terminology, and patient care skills. I especially enjoyed assisting the medical professionals in sometimes highly stressful situations and helping patients and their family. I would enjoy contributing to Dr. Bewell's success serving as a medical assistant.
Culinary	For the past five years, I worked in the food service industry, where I learned the importance of accuracy, quality, and customer service. I actually started as a fast food cook at a local grill and decided I wanted to expand my culinary skills. I am working toward a degree in culinary arts and am quickly discovering that I particularly enjoy the baking/pastry activities. I've gained extensive experience in food safety, inventory control, and creating decorative food displays. I am hardworking, creative, and resourceful, and I play the tuba in my spare time. Thank you for this opportunity to meet with you. I promise, you won't regret hiring me.

Use your personal commercial at the start of your interview when asked, "Tell me about yourself." If you are not given this instruction during the interview, include it at the end. Memorize your personal commercial and, in front of a mirror, practice delivering it until it is communicated naturally and confidently.

The Interview Portfolio

An **interview portfolio** contains relevant documents that are taken to an interview. Use a professional business portfolio or a high-quality paper folder with pockets. The portfolio should be a color that will not draw attention. Include copies of items pertinent to the position for which you are applying. Original documents (unless required) should not be given to the employer, only photocopies. Have the following items in your interview portfolio: copies of your résumé, cover letter, reference list, generic application, and a copy of your personal commercial. Also include a calendar, note paper, a pen, and personal business cards. Print copies of your résumé, cover letter, and references on résumé paper. Copies of other items such as skill or education certificates, recent performance evaluations, job samples, or other documents you have in your job search portfolio may be included if the information is relevant to the job. During the interview keep your interview portfolio on your lap. Place your personal commercial on the top of your portfolio for easy access. Do not read the commercial. Use the copy for reference or to merely glance at, should you become nervous and forget what to say. Use the checklist in Table 15.1 to ensure the items are included in your interview portfolio. Additionally, ensure these items are included in your electronic portfolio (e-portfolio) created in Chapter 13.

Table 15.1	Interview Portfolio	
Item	**Description**	**Included**
Copies Necessary for Each Interview		
Awards	Educational, professional, and personal	
Calendar (electronic or small traditional)	To track dates	
Completed generic application	Available information if asked to fill out an application on site	
Cover letter	Signed and on résumé paper	
Letters of recommendation	No older than one year	
Note pad	To document names and notes from interview	
Pen	For note taking	
Performance appraisals from previous jobs	Proof of positive work performance	
Personal business cards	Share with the interviewer	
Personal commercial	Use to glance for reference	
Reference list	On résumé paper	
Résumé	On résumé paper	
Transcripts	A copy of official school transcripts	
Copies if Relevant for the Job		
Certificates	To prove official relevant certification	
Copy of ID and/or driver's license	If driving is a job requirement	
Copy of recent DMV record (if relevant to your career)	If driving is a job requirement	
Current state licenses	To prove current relevant licensing	
Work samples	Examples to demonstrate proficiency in specific skills	

Practice Interview Questions

Another activity to perform when preparing for an interview is to practice interview questions. Table 15.2 identifies common interview questions, appropriate answers, and topics to avoid. Review this list and begin creating appropriate responses to each question. When answering an interview question, avoid simply answering "yes" or "no" or providing a generic answer. Elaborate your response by providing examples of specific skills and experiences that support your answers and meet the key requirements of the target job. These experiences do not have to be based on work experience; you may use examples from school or volunteer work. The more real-life examples you provide, the more you demonstrate your experience and skill level to the employer. Anyone can say, "I can handle stress." However, providing a specific example of how you handled stress on a busy day clearly communicates how you handle stress.

With each answer you are being judged on your qualifications and how you will fit into the culture of the company. Interviewers look for what value you will bring to the company. Listen to each question carefully and determine what the employer is trying to identify about you. If you are asked a question you do not understand, do not ask the interviewer to repeat the question, as the interviewer may think you were not listening. Ask for clarification. Although some questions can be answered with a simple yes or no or a basic statement, fully answer the question being asked by presenting your related skill and/or experience. For example, if the interview asks, "How many years of experience do you have working in a law office?" do not simply respond with a number. Elaborate by sharing what type of legal environments you have worked in and what type of activities you performed.

With any question asked, be prepared to provide at least two specific examples. Use personal experiences from a previous job, an internship, classwork, or a school-related activity. If the question involves sharing a challenge or hardship you have faced, avoid providing personal details and use examples that demonstrate your ability to overcome a challenge. For example, if you are asked how you handle stressful situations and you have no work experience, share a time you worked on a team project in school and you had a member not performing his portion of the project. Although the team member was making you angry and causing you stress, mention how you used stress methods (from Chapter 3) and state how you took control of the situation by displaying leadership and constantly communicated with the team member to hold him accountable.

Table 15.2　Common Interview Questions

Question	Response	Avoid
Tell me about yourself.	Use your personal commercial modified to the job description.	Do not divulge where you were born, your age, if you are married, if you have children, or other personal information.
What are your strengths?	Include how your strengths meet the job requirements and how they will be an asset to the company.	Do not include strengths that are not related to the job. Do not include personal information (e.g., "I'm a good mother").

Tell me about a time you failed.	Use an example that is not too damaging. Turn it into a positive by including the lesson learned from the experience.	Do not place blame for why the failure occurred.
Tell me about a time you were successful.	Use an example that relates to the job for which you are applying.	Avoid coming across as arrogant. Do not take full credit if the success was a team effort.
How do you handle conflict?	Use an example that is not damaging and discuss how the conflict was positively resolved. (Addressed in Chapter 12)	Avoid examples that place you in a negative light. Do not provide specifics on how the conflict occurred, and do not place blame on others.
Would you rather work individually or in a team? Why?	State why you prefer one or the other, but relate your answer to the job requirements.	Do not state that you will not work one way or the other.
Why do you want this job?	Convey career goals and how the specific job/company supports your current skills. Include company information learned through research.	Do not include money or benefits in your response.
How do you deal with stress?	Share positive stress reducers (addressed in Chapter 3).	Do not state that stress does not affect you. Do not use negative examples.
What is your greatest weakness?	Use a weakness that will not damage your chance of getting the job. Explain how you are minimizing your weakness or are turning it into strength (e.g., "I have not been particularly organized in the past; but since going to school I have used my lessons in time management and improved").	Do not state, "I don't have any." Do not just state a weakness; instead, show how you are turning it into a strength.
Where do you want to be in five years?	Share the career goals you created in Chapter 1.	Do not say you want the interviewer's job.
Tell me about a time you displayed leadership.	Use a specific example and try to relate it to the needed job skills.	Do not appear arrogant.
If you are currently employed, why do you want to leave your job?	Be positive and specific. Relate back to education and career goals. Be honest if you are losing your job based on a change in company strategy.	Do not speak poorly of a company or its employees.
Why do you want to work for us?	Share the company information from your research and relate it to your career goals.	Do not include money or benefits in your response.
What motivates you?	Identify tasks you have completed that motivated you to finish a project.	Do not include money or benefits in your response.
What was your favorite subject in school?	Identify a subject that is related to the target job and add why you enjoyed the course. If you have a secondary subject that is unique (e.g., athletics or music), share that as well.	Do not say "all of them" or "none of them."
What are your salary expectations?	Based on your research, provide a salary range. Avoid discussing specific compensation issues until the final stages of the hiring process.	Do not provide a specific number. When it comes time to discuss salary, identify the lowest and ideal salary so you have room to negotiate.

Practice answering interview questions in front of a mirror, and, if possible, create a video of yourself answering common interview questions. Critically analyze your responses to see if you are appropriately answering the questions, selling your key skills, and projecting a professional image. You should appear honest, confident, and sincere. If you display nervous gestures, work on eliminating the distracting gestures. This exercise will better prepare you for an interview and will increase your self-confidence.

Pre-Interview Preparation

Prior to the day of your interview, conduct a "practice day" interview. If possible, travel to the interview location, ideally during the same hour as your scheduled interview to identify potential transportation problems such as traffic and parking. Once you arrive at the site, walk to the specific location where the interview will be held. This will enable you to become comfortable and familiar with your surroundings and let you know how much time you will need to arrive on time the day of the interview. Do not go into the specific office, just the general area. Make note of the nearest public restroom so you can use it the day of the interview to freshen up prior to your meeting.

Prior to the day of the interview, also ensure that your interview attire is clean and you have a professional appearance. Dress at a level above the position for which you are interviewing. For example, if you are interviewing for an entry-level position, dress like you are interviewing for a supervisor position. Clothes should fit appropriately and shoes should be clean. Cleanliness is important. Ensure that your hair and fingernails are professional and appropriate for an interview. If necessary, get a haircut prior to your interview. Use little or no perfume/aftershave and keep jewelry to a minimum. When conducting research, identify the company's dress code policy. Depending on the job and industry, you may need to cover tattoos and remove extra piercings (if you have them). Refer to Chapter 4 to review professional dress in greater detail.

Customize your interview portfolio by including copies of any documents related to the target job. Place your portfolio where you will not forget it when you leave your home. With the common use of smartphones and other portable electronic devices, it is a good idea to have your e-portfolio items ready for use in the interview. If you decide to utilize your e-portfolio, inform the interviewer that you would like to use your device during the interview to show him or her items from your e-portfolio.

Purchase a package of simple but professional thank-you notes. The evening before your interview, write a draft post-interview thank-you note on a blank piece of paper. Keep the thank-you note brief, only three to four sentences. In the note, thank the interviewer for his or her time. State that you enjoyed learning more about the position, are very interested in the job, and look forward to hearing from the interviewer soon. The draft note will be used as a foundation for notes (traditional or electronic) you will be writing and sending immediately after your interview. Place the draft note, the package of thank-you notes, several personal business cards, and a black pen alongside your interview portfolio to take with you.

Exercise 15.3

Write a draft post-interview thank-you note.

_____

_____

_____

The Interview Process

The interview process varies from employer to employer. One position may only require a one-on-one interview and then a job offer, whereas another position may involve pre-employment testing, a phone interview with the human resource department, then a panel interview with potential coworkers, and then a face-to-face interview with the department manager. Some hiring processes will occur rather quickly; other employer hiring processes may take months. The following sections will explain in greater detail the various venues, methods, and types of interviews.

Interview Day

On the day of the interview be well rested and have food in your stomach prior to the interview. Look in the mirror to ensure a professional appearance. Clothes should fit properly and project a professional image. If you smoke, refrain from smoking prior to the interview. The smell may be a distraction to the interviewer.

Be on time and plan to arrive at your destination 15 minutes early. This provides time to deal with unforeseen transportation issues. If there is a public restroom available, visit the restroom and freshen up. Check your hair, clothing, and makeup, if applicable. Turn off your communication device, and if you are chewing gum, throw it away. Enter the specific meeting location 5 but no more than 10 minutes prior to your scheduled interview.

This is where your interview unofficially begins. First impressions matter and any interaction with representatives of the organization must be professional.

Immediately upon entering the interview location, introduce yourself to the receptionist. Offer a smile and a handshake, and then clearly and slowly state your name. For example, "Hi, I'm Cory Kringle, and I am here for a 9:00 A.M. interview with Ms. Dancey for the accounting clerk position." If you recognize the receptionist as the same individual who arranged your interview appointment, make an additional statement thanking the individual for his or her assistance. For example, "Mrs. Wong, were you the one that I spoke with on the phone? Thank you for your help in arranging my interview." Be sincere in your conversation, and convey to the receptionist that you truly appreciate his or her efforts. The receptionist will most likely ask you to have a seat and wait to be called into the interview. Take a seat and relax. Avoid using your mobile device during this time. While you are waiting, use **positive self-talk**. Positive self-talk is a mental form of positive self-reinforcement. It helps remind you that you are

Talk It Out

Why should you not arrive too early to the meeting location?

qualified and deserve both the interview and the job. Mentally tell yourself that you are prepared, qualified, and ready for a successful interview. Review your personal commercial, your qualifications, and the key skills you want to convey in the interview.

Topic Situation

Johnelle's friend Shelby had been asked to interview with one of her target companies. Shelby really wanted the job but was afraid she was not going to do well during her interview. Johnelle worked with Shelby a few days before the interview by role-playing interview questions and reviewing Shelby's company research. Johnelle had Shelby dress in her interview attire. They then videotaped a mock interview and critiqued the video. Finally, they traveled to the interview site and made sure Shelby knew the exact interview location. The next day, when Shelby arrived for the interview, she arrived early, thanked the receptionist, and took a seat. As she waited to be called in to the interview, she began getting extremely nervous. Remembering Johnelle's tips, Shelby briefly closed her eyes and used positive self-talk to improve her attitude, increase her confidence, and calm her nerves. After doing this, she felt more confident when called in to the office to begin the interview.

TOPIC RESPONSE

What more could Shelby and Johnelle have done to prepare for the interview?

Traditional Face-to-Face Interviews

A traditional interview involves a face-to-face meeting between an applicant and the employer. As with any type of interview, convey confidence. Your primary message during the interview will be how your knowledge, skills, and abilities will be assets to the company and make you the best candidate for the job. When you are called to interview, stand up straight and approach the individual who called your name. If it is not the receptionist who called you, extend a smile and a hand-shake, and then (speaking clearly and slowly) introduce yourself. For example, "Hi, I'm Cory Kringle. It's nice to meet you." Listen carefully to the interviewer's name so you will remember it and use it during the interview. He or she will escort you to an office or conference room where the interview will take place. If you enter a room and there is someone in the room that you have not met, smile, walk to the individual, extend a handshake, and introduce yourself. Once in the room, do not be seated until you are invited to do so. When seated, write down the names of the individuals you have just met. Whenever possible, interject the inter-viewer's name(s) during the interview. Although you may be offered something to drink, it is best to decline the offer so there is nothing to distract you from the interview. If you are sitting in a chair that swivels, place your feet flat on the floor to remind yourself not to swivel. If you forgot to turn off your communication device and it rings during the interview, do not answer it. Immediately apologize to the employer and turn it off. Do not take time to see who called.

If the interview is taking place in an office, look around the room to get a sense of the person who is conducting the interview, assuming it is his or her office. Doing so provides useful information for conversation, should it be nec-essary. Depending on the time available and the skills of the interviewer(s), you may first be asked general questions, such as, "Did you have trouble finding our office?" The interviewer is trying to get you to relax. During the interview, pay attention to body language—both yours and that of the individual conducting the interview. Sit up straight, sit back in your chair, and try to relax. Be calm and confident, but alert. Keep your hands folded on your lap or ready to take notes, depending on the situation. If you are seated near a desk or table, do not lean on the furniture. Make eye contact, but do not stare at the interviewer. Your job is to connect with the interviewer by being sincere and personable.

If you are given the opportunity to provide an opening statement, share your personal commercial. If you are not able to open with your personal commercial, include it in an appropriate response or use it at the end of the interview. Do not talk over the interviewer or interrupt when he or she is speaking. When asked a question, listen carefully. Take a few seconds to think about what information the interviewer truly wants to know. Formulate an answer and relate your response back to the job qualifications and/or job duties. Avoid cliché or general answers. As covered in a previous section, your goal is to favorably stand out from other candidates while conveying how your skills will assist the company in achieving success. Provide a complete response without rambling. Sell your skills and expertise by including a specific example, and whenever possible, interject information you learned about the company during your research.

As shared earlier, employer interviewer styles and questions differ. In every situation, demonstrate how you are the best match for the target job. If you are asked questions that appear completely irrelevant to the interview (e.g., "If you were invited to a department potluck, what item would you bring and why?"), do not get defensive. Remain professional and be sincere in your response. When interviewers utilize silence, they are most likely gaining an understanding of how you handle stressful situations. Throughout the interview, be aware of and avoid nervous gestures. Remain relaxed, poised, and confident.

Interview Methods and Types of Interview Questions

There are several common interview methods, including one-on-one interviews, group interviews, and panel interviews. **One-on-one interviews** involve a one-on-one meeting between the applicant and a company representative. The company representative is typically either someone from the human resource department or the immediate supervisor of the department with the open position. **Group interviews** involve several applicants interviewing at the same time while being observed by company representatives. The purpose of a group interview is to gauge how an individual behaves in a competitive and stressful environment. In a group interview, practice positive human relation and communication skills toward other applicants. Listening and communicating that you are the best candidate is critical to a successful group interview. If another applicant is first asked a question and you are immediately asked the same question, do not repeat what the other applicant said. If you agree with the first applicant's response, state, "I agree with Ms. Bell's response and would like to add that it is also important to . . . ," and then elaborate or expand on the first applicant's response. If you do not agree with the first applicant's response, state, "I believe . . . ," and then confidently provide your response. Show respect by not demeaning other applicants. Be professional, do not interrupt, and behave like a leader by being assertive, not aggressive.

Panel interviews involve the applicant meeting with several company representatives at the same time. During a panel interview, make initial eye contact with the person asking the question. While answering the question, connect with all members of the interview panel. Whenever possible, call individuals by name. As with a one-on-one interview, your job is to appear personable and attempt to create a favorable impression that makes each panelist want to hire you.

The three general types of interview questions are structured, unstructured, and behavioral. **Structured interview questions** address job-related issues where each applicant is asked the same question(s). The purpose of a structured interview question is to secure information related to a specific job. An example of a structured question is, "How long have you worked in the retail industry?" Although the sample question appears to be closed-ended, a skilled candidate will elaborate on his or her response. For example, avoid simply answering, "I have worked in retail for two years." Add additional information to your response; for example, "I have two years of retail experience. One year I worked in a small family business and the other for a large retail chain. During this time I have increased my general knowledge of business and my customer service skills. I liked the family business best because I enjoyed getting to know our loyal customers."

An **unstructured interview question** is a probing, open-ended question. The purpose of an unstructured interview question is to identify if a candidate can appropriately sell his or her skills. An example of an unstructured interview question is, "Tell me about yourself." When you are asked to talk about yourself, start with your personal commercial and, if appropriate, show job samples from your interview portfolio when referring to a specific skill. Make every attempt to provide specific examples without rambling and relate answers back to the target job.

Behavioral interview questions are questions that ask candidates to share a past experience related to a workplace situation. The purpose of a behavior interview is to identify what a candidate has done in the past, including how the candidate behaves under a specific circumstance. An example of a behavioral question is, "Describe a time you motivated others." Prior to answering the question, take a moment to formulate your answer. Attempt to have the interviewer view the situation from your perspective by briefly providing the background to your experience. Describe how you used specific skills to solve a problem or improve a situation. For example, "Our department had just gone through a series of layoffs and employees were working hard but feeling unappreciated. Although I was not the team leader, I thought we needed something to help us deal with the stress, so for one week, and with my direct report's permission, I planned brief, fun activities for our daily morning meetings. The first day, we played a game I made up called 'Name That Stress.' At first glance, it seemed silly, but it actually started a conversation about how stressed we all were and how we could collectively deal with the stress in a challenging situation."

Phone and Other Technology-Based Interviews

In some situations, your first interview may take place over the phone. Phone interviews may or may not be prearranged. During your job search, consistently answer your phone in a professional manner and keep your interview portfolio in an accessible place. If a company calls and asks if it is a good time to speak with you and it is not, politely respond that it is not a convenient time and ask if you can reschedule the call. Try to be as accommodating as possible to the interviewer.

Those being interviewed by phone should follow these tips:

- *Be professional and prepared.* Conduct the interview in a quiet room and focus solely on the conversation. Unless an electronic device (e.g., computer) is necessary for part of the interview, turn it off. Remove any additional distractions, including music, pets, television, and other individuals from your quiet area. Company research, personal examples, and the use of your personal commercial are just as important to interject into the phone conversation as during a face-to-face interview. Take notes and ask questions.
- *Be concise with your communication.* Those conducting the phone interview are not able to see you; therefore, they are forming an impression of you by what you say and how it is stated. Speak clearly, at a normal pace, and do not interrupt. When responding, count to three in your head to allow time for telecommunications overlap. Speak naturally, but loud enough for the interviewer to hear and understand you. Smile and speak with enthusiasm throughout the interview. Use proper grammar and beware of "ums" and other nervous verbal phrases. If you stand while conducting your phone interview, you will stay alert, focused, and more aware of your responses.
- *Be polite.* Utilize what you learned in both the etiquette and communication chapters. Exercise good manners. Do not eat or chew gum during your interview. It is not appropriate to use a speaker phone when being interviewed, nor is it polite to take another call or tend to personal matters. Your attention should be completely focused on the interview. When the conversation is over, ask for the job, and thank the interviewer for his or her time.

It is becoming increasingly common for interviews to take place through video chat venues such as Skype, WebEx, and Google Talk. An individual participating in a video chat interview needs a computer, a web cam, and a reliable Internet connection. When taking part in a video chat interview, the participant will receive a designated time and specific instructions on where and how to establish the connection. In addition to following the phone interview tips, the interviewer needs to prepare and treat the video chat interview as if it were a face-to-face interview. Therefore, use the following tips:

- *Plan ahead.* Research the venue you will be using to address any unforeseen issues. Identify where you will conduct the interview and what technology is required. If possible, arrange a pre-interview trial to ensure all equipment works properly and you know how to use it (including the volume and microphone).
- *Dress professionally.* Attire should be the same as a face-to-face interview. You will be in plain view of the interviewer, so visual impressions matter.
- *Maintain a professional environment.* Conduct your interview in a quiet and appropriate location void of distractions. Ensure the background is appropriate. A bedroom, public place, or outside location is not appropriate.
- *Speak to the camera.* Focus on the web cam as if it were the interviewer's face. Feel free to ask questions, take notes, and use hand gestures. Although it may be more difficult to communicate, make every effort to not only project your personality, but, more importantly, sell your knowledge, skills, abilities, and unique qualifications. As with a traditional face-to-face interview, your job is to connect with the interviewer.

Additional information regarding technology-based meetings is contained in Chapter 10.

Discrimination and Employee Rights

Title VII of the Civil Rights Act was created to protect the rights of employees. It prohibits employment discrimination based on race, color, religion, sex, or national origin. Other federal laws prohibit pay inequity and discrimination against individuals 40 years or older, individuals with disabilities, and individuals who are pregnant. This does not mean that an employer must hire you if you are a minority, pregnant, 40 or older, or have a disability. Employers have a legal obligation to provide every qualified candidate equal opportunity to interview. Their job is to hire the most qualified candidate. Although discriminatory questions are illegal, unfortunately, some employers may ask them in an interview. Table 15.3 was taken from the California Department of Fair Employment and Housing to provide examples of acceptable and unacceptable employment inquiries.

If an interviewer asks you a question that is illegal or could be discriminatory, do not directly answer the question. Instead, answer the question in

Table 15.3   Illegal Interview Questions

Subject	Acceptable	Unacceptable
Name	Name	Maiden name
Residence	Place of residence	Questions regarding owning or renting
Age	Statements that employment is subject to verification if applicant meets legal age requirement	Age Birthdate Date of attendance/completion of school Questions that tend to identify applicants over 40
Birthplace, citizenship	Statements/inquiries regarding verification of legal right to work in the United States	Birthplace of applicant or applicant's parents, spouse, or other relatives Requirements that applicant produce naturalization or alien card prior to employment
National origin	Languages applicant reads, speaks, or writes, if use of language other than English is relevant to the job for which applicant is applying	Questions as to nationality, lineage, ancestry, national origin, descent, or parentage of applicant or applicant's spouse, parent, or relative
Religion	Statement by employer of regular days, hours, or shifts to be worked	Questions regarding applicant's religion Religious days observed
Sex, marital status, family	Name and address of parent or guardian, if applicant is a minor Statement of company policy regarding work assignment of employees who are related	Questions to indicate applicant's sex, marital status, number/ages of children or dependents Questions regarding pregnancy, child birth, or birth control Name/address of relative, spouse, or children of adult applicant
Arrest, criminal record	Job-related questions about convictions, except those convictions that have been sealed, expunged, or statutorily eradicated	General questions regarding arrest record

Source: California Department of Fair Employment and Housing Fact Sheet. DFEH-161 (8/01)

an indirect manner by addressing the employment issue and stating how your qualifications are well-suited for the job. For example, if the interviewer says, "You look Hispanic. Are you?" your response should not be "Yes" or "No." Politely smile and say, "People wonder about my ethnicity. What can I tell you about my qualifications for this job?" Do not accuse the interviewer of asking an illegal question or say, "I will not answer that question because it is illegal." Most employers do not realize they are asking illegal questions. However, some employers purposely ask inappropriate questions. If you are asked several illegal questions, you need to decide if you want to work for an employer who either does not properly train its interviewers or intentionally asks illegal questions.

Know and protect your rights. It is inappropriate to disclose personal information about yourself during an interview. Avoid making any comment referring to your marital status, children, religion, age, or any other private issue protected by law.

Talk It Out

Role-play an interview. During the interview, ask one legal question and one illegal question. Practice answering the illegal question with confidence but in a non-offensive manner.

Special Circumstances and Tough Questions

Life is unpredictable and sometimes results in situations that can be embarrassing or difficult to explain during a job interview. These situations may include a negative work experience with a previous employer, time gaps in a résumé, or a prior felony conviction. The following information provides the proper response to interview questions related to these difficult situations.

Some job seekers have had negative work-related experiences that they do not want to disclose during an interview. Disclosing such information could be potentially devastating to a job interview if it is not handled properly. Some of these experiences include being fired, quitting due to poor working conditions, having a poor performance evaluation, or knowing that a former manager or teacher will not provide a positive reference if called. Perhaps you behaved in a negative manner prior to leaving your old job.

If you had a negative work experience and are not asked about the situation, there is no need to disclose the unpleasant event. The only exception to this rule is if your current or former boss has the potential to provide a negative reference. If this is the situation, tell the interviewer that you know you will not receive a positive reference from him or her and request that the interviewer contact another manager or coworker who will provide a fair assessment of your performance.

Being honest and factual is the best answer to any difficult question. If you were fired, performed poorly, or left in a negative manner, state the facts, but do not go into great detail. Tell the interviewer that you have matured and realize that you did not handle the situation appropriately. Add what lesson you have learned. Do not speak poorly of your current or previous employer, boss, or coworker. Also avoid placing blame by stating who you feel was right or wrong in the negative workplace situation.

As stated in Chapter 14, it is common for an individual to have time gaps in a résumé as a result of an extended job search, staying home to raise a young child, caring for an elderly relative, or continuing his or her education. Those who have gaps in their résumé may need to be prepared to explain what they did during the time gap. Identify a key skill you sharpened during your time gap and relate this experience to a key skill necessary for your target job and industry. If you volunteered for projects, include this information. For example, if you stayed at home to care for an elderly relative and are asked about the time gap, explain the situation without providing specific details, and then share how the experience improved your time management and organizational skills in addition to improving your awareness of diverse populations, including the elderly and disabled.

If you have a felony record, you may be asked about your conviction. As with other difficult interview questions, be honest and factual in your response. Explain the situation and tell the interviewer that you have made restitution, are making every attempt to start anew, and are committed to doing your very best. Sell your strengths and remember to communicate how your skills will help the company achieve its goals. Your self-confidence and honesty will be revealed through your body language and eye contact. Be sincere. Depending on the type and severity of your offense, it may take more attempts to secure a job than during a typical job search. You may also need to start at a lower level and/or lower pay than desired. The goal is to begin to reestablish credibility. Do not give up. Each experience, be it positive or negative, is a learning experience.

Closing the Interview

After the interviewer has completed his or her questioning, you may be asked if you have any questions. Having a question or closing statement prepared for the close of your interview demonstrates to the prospective employer that you have properly prepared for the interview. A good question refers to a current event that has occurred within the company. For example, "Ms. Dancey, I read about how your company employees donated time to clean up the ABC school yard. Is this an annual event?" A statement such as this provides you one last opportunity to personalize the interview and demonstrate that you researched the company. This is also a good time to share additional relevant information you have in your portfolio that you were not able to present during the interview.

Do not ask questions that imply you did not research the company or that you care only about your needs. Inappropriate questions include questions regarding salary, benefits, or vacations. These questions imply that you care more about what the company can do for you than what you can do for the company. However, it is appropriate and important to ask what the next steps will be in the interview process, including when a hiring decision will be made. Table 15.4 contains questions that may and may not be asked.

Table 15.4 Closing Interview Questions	
Questions You May Ask	**Questions You Should Not Ask**
What is the next step in the interview process?	How much does this job pay?
What do you enjoy about working for this company?	What benefits will I get?
What type of formal training does your company provide?	How many vacation days do I get?
What are you looking for in an employee?	How many sick days do I get?
Does your company have any plans for expansion?	What does your company do?
What is the greatest challenge your industry is currently facing?	How long does it take to get a raise?

After the interviewer answers your general questions, make a closing statement. Summarize your personal commercial and ask for the job. An example of a strong closing statement is: "Once again, thank you for providing me the opportunity to interview, Ms. Dancey. As I stated at the beginning of our meeting, I feel I am qualified for this job based on my two years of retail experience, business knowledge, and demonstrated leadership. I would like this job and believe I will be an asset to XYZ Company." The purpose of a job interview is to sell yourself and your skills. A sale is useless if you do not close the sale by asking for the job.

Exercise 15.4

Write a closing interview statement.

_____

_____

_____

After you make your closing statement, the interviewer will signal that the interview is over. He or she will do this either through conversation or through body language, such as standing up and walking toward the door. Prior to leaving the interview, hand the interviewer your personal business card and ask the interviewer for a business card. You will use this business card for the interview follow-up. Look your interviewer in the eye, shake the interviewer's hand, thank him or her, and state that you look forward to hearing from him or her soon. Continue communicating confidence, friendliness, and professionalism to every employee you encounter on your way out of the building.

After you have left the building, retrieve your draft thank-you note. If appropriate, modify the draft thank-you note to include information that was shared during your interview. Handwrite a personalized thank-you note to each individual who interviewed you. Use your finest handwriting and double-check spelling and grammar. Refer to the business card(s) you collected for correct name spelling. Do not forget to sign the thank-you note and include a personal business card if you have one. After you have written your note, immediately hand deliver it to the reception area and ask the receptionist to deliver the notes. Your goal is to make a positive last impression and stand out from the other candidates. If you are unable to provide a handwritten note immediately after the interview, send an e-mail thanking the company representatives the same day as the interview. As with a handwritten thank-you note, ensure your communication is professional, error-free, and includes identifying contact information.

After the Interview

After sending or delivering your thank-you notes, congratulate yourself. If you did your best, you should have no regrets. When you are able, make notes regarding specific information you learned about your prospective job and questions you were asked during the interview. In the excitement of an interview, you may forget parts of your meeting if you do not immediately make notes. Write down what you did right and areas in which you would like to improve. This is a good time for you to evaluate your impressions of the company and determine if it is a company where you will want to work. This information will be helpful in future interviews.

Salary Negotiation

Soon after your initial interview, you should hear back from the company. At that point, you may be called in for a second interview or may receive a job offer. A job offer may be contingent upon reference and background checks. This will be a good time to contact the individuals on your reference list to provide them an update on your job search and ensure they are prepared to respond appropriately to the individual conducting your reference check.

If you are a final candidate for the job, the interviewer may ask you about your salary requirements. In order to negotiate an acceptable salary, first conduct online research and compare your research to the salary range that was included in the job announcement. Determine local and regional salaries and attempt to match the job description as closely as possible to that of the job for which you are applying. Depending on your experience, start a few thousand dollars higher than your desired starting salary and do not forget to consider your experience or lack thereof. Some companies do not offer many benefits but offer higher salaries. Other companies offer lower salaries but better benefits. Weigh these factors when determining your desired salary. Prior to stating your salary requirement, sell your skills. For example, "Ms. Dancey, as I mentioned in my initial interview, I have over five years' experience working in a professional accounting office and an accounting degree; therefore, I feel I should earn between $35,000 and $45,000." If you are offered a salary that is not acceptable, use silence and wait for the interviewer to respond. This minute of silence may encourage the employer to offer a higher salary.

Topic Situation

TOPIC RESPONSE

Why is it important to research, know, and state a desired salary when asked during an interview?

Isaac had a successful interview because he did his research and practiced questions. Isaac received a call from the employer and was invited to a second interview. Prior to the interview, Isaac reviewed notes from his first interview and again prepared for potential questions and situations that he might encounter. Isaac had his friend Bret help him practice for the second interview. In their practice, Bret asked Isaac about his starting salary. Isaac said he did not care; he would just be happy to get a job. Bret reminded Isaac that he needed to sell his skills and have a desired target salary. They then conducted an Internet search of both local and statewide jobs that were similar to the one Isaac wanted.

Pre-Employment Tests, Screenings, and Medical Exams

Pre-employment tests are assessments that are given to potential employees as a means of determining if the applicant possesses the desired knowledge, skills, or abilities required for a job. Pre-employment tests can be giving during the application process, during the interview process, or prior to receiving a job offer. Some employers require applicants to take online pre-employment tests. Some tests may require lifting, others are skills-based, and others measure listening or logic. Legally, pre-employment tests must be job-related. Depending on the type of test, you may be given the results immediately. In other cases, you may need to wait for the results. If you pass the employment test(s), you will be invited to proceed with the interview process. It is common for employers to have applicants who did not pass a pre-employment test to wait a predetermined period prior to reapplying.

Employers may also conduct pre-employment screenings and medical exams. Pre-employment screenings include criminal checks, education verification, driver's license history, security checks, previous employment checks, credit checks, reference checks, and drug tests. The number and type of pre-employment screenings performed will be based on how relevant the check is to the job you will be performing. Legally, employers can require medical exams only after a job offer is made, but they may require pre-employment drug tests. The exam must be required for all applicants for the same job, and the exam must be job-related. Employers are not allowed to ask disability questions related to pre-employment screenings and medical exams. Common medical exams include vision and strength testing.

An employer legally cannot conduct these checks without your permission. Most employers will secure your permission in writing when you complete an employment application or when you are a finalist for the position.

When You are Not Offered the Job

A job search is a full-time job and can sometimes be discouraging. When you are not invited to interview, reevaluate your job search approach and tools. First consider your target companies. Ensure you are applying for jobs that fit your current knowledge, skills, and abilities. Continuous networking increases the chances your résumé package will receive the proper attention. Pay attention to application instructions and deadlines. Only submit complete application materials and submit them on time. Most companies require a completed application, cover letter, and résumé. Some companies require all submissions be made online, whereas others prefer a traditional approach with a simple paper résumé package. Evaluate your résumé and cover letter. Your cover letter should spark curiosity in the reader by emphasizing select details of your qualifications and make them want to read the attached résumé. Review both for typographical or grammatical errors. Ensure you have referenced key words directly from the job description and that your résumé lists important skills that reflect the needs of your target job. Have someone who knows you and your skills, and whom you trust, review your cover letter and résumé. Many times, a fresh perspective will catch obvious errors or opportunities for improvement. Although many job applicants state they will follow up in their cover letter, many do not. If you indicated you would follow up, do so within two weeks of submitting your application package. This provides you an additional opportunity to communicate your interest and qualifications for the target position.

If you are invited to interview but do not receive a job offer, do not be discouraged. Remember to make every experience a learning experience. Review your personal commercial to ensure it is personal, powerful, and unique. With notes you took from past interviews, carefully review each step in the interview process and grade yourself. Consider your preparation, your appearance, your responses, your ability to interject company research into each answer, and your overall attitude. Any area that did not receive an "A" grade is an area poised for improvement.

There are several steps you can take to increase the probability for success in your next interview. Consider your overall appearance. Reviewing the information in Chapter 4, make sure you convey professionalism from head to toe. Ensure that your clothes are clean and fit properly. Have a hairstyle that is flattering and well-kempt. Check that your fingernails and jewelry are appropriate and do not distract from your personality and job skills.

Mentally review job interview questions that were asked and the responses you provided. Every answer should communicate how your knowledge, skills, and abilities will assist the target company in achieving success. Review the amount of company research you conducted. Did you feel amply prepared, or did you simply research the bare minimum? If you felt you did conduct the appropriate amount of research, evaluate whether you fully communicated your research to the interviewer.

Assess your body language and attitude. Stand in front of a mirror and practice your answers to difficult and/or illegal questions. If possible, have a friend videotape you and provide an honest evaluation of your appearance, attitude, and body language. Check for nervous gestures, and keep practicing until you are able to control nervous habits.

Be honest about your overall performance before, during, and after the interview. Review your activities after the interview, including immediately sending a thank-you note. Prior to your next interview, identify at least two areas of the interview process you would like to improve and begin working on those areas. Improving these areas will make you a stronger interviewee so that you stand out above all other candidates and receive a coveted job offer.

MyStudentSuccessLab Please visit **MyStudentSuccessLab**: Anderson|Bolt, Professionalism Skills for Workplace Success, 4/e for additional activities, resources, and outcomes assessments.

Workplace Dos and Don'ts

Do tailor your résumé and personal commercial to the needs of your targeted employer	*Don't* have unprofessional introductions on your voice mail message
Do try to schedule your interview at a time that puts you at an advantage over the other candidates and secure information that better prepares you for the interview	*Don't* make demands from the individual scheduling the interview
Do learn as much as you can about the company, its strategy, and its competition	*Don't* arrive late or too early
Do practice interview questions and formulate answers that highlight your skills and experience	*Don't* forget to include company and industry research information in your interview responses
Do remember that your interview begins the minute you step onto company property	*Don't* show up to an interview unprepared and empty handed
Do know how to handle inappropriate questions that may be discriminatory	*Don't* let your nerves get the better of you in a job interview
Do prepare questions to ask the interviewer	*Don't* answer an illegal question; instead address the issue

Concept Review and Application

You are a Successful Student if you:

- Create an interview portfolio
- Videotape yourself in a practice interview
- Identify and create responses to difficult interview questions specific to you and your situation

Summary of Key Concepts

- Create and modify your personal commercial and adapt it to the requirements of your target job.

- Review common interview questions and formulate answers as part of your interview preparation.

- Conduct a pre-interview rehearsal to ensure you are prepared the day of the interview.

- During your interview, communicate how your knowledge, skills, and abilities will be assets to the company.

- Understand the laws that protect employees from discrimination in the interviewing and hiring process.

- Be prepared to confidently handle gaps in employment and other difficult interview questions.

- Know how to sell yourself and professionally ask for the job at the close of an interview.

Self-Quiz MATCHING KEY TERMS

Match the key term to the definition using the identifying number.

Key Terms	Answer	Definitions
Behavioral interview question		1. An interview that involves a one-on-one meeting between the applicant and a company representative
Group interview		2. A type of interview question that addresses job-related issues where each applicant is asked the same question
Interview portfolio		3. A mental form of positive self-reinforcement that helps remind you that you are qualified and deserve both the interview and the job
One-on-one interview		4. An interview that involves several applicants interviewing at the same time while being observed by company representatives
Panel interview		5. A brief career biography that conveys one's career choice, knowledge, skills, strengths, abilities, and experiences
Personal commercial		6. An interview that involves the applicant meeting with several company employees at the same time
Positive self-talk		7. A probing, open-ended interview question intended to identify if the candidate can appropriately sell his or her skills
Structured interview question		8. Interview question that asks candidates to share a past experience related to a specific workplace situation
Unstructured interview question		9. A folder to be taken on an interview that contains photocopies of documents and items pertinent to a position

Think Like a Boss

1. What kind of information should you share with your current staff members as they prepare to interview a new employee?

2. How would you handle a prospective employee who disclosed inappropriate information during the job interview?

Activities

Activity 15.1

Identify a local company for which you would like to interview. Using the following table, conduct a thorough targeted job search on this company. Answer as many of the questions as possible.

Company name	
Company address	
Job title	
To whom should the cover letter be addressed?	
What are the job requirements?	
Is this a full-time or part-time job?	
What are the hours/days of work?	
What are the working conditions?	
Is there room for advancement?	
What kind of training is offered?	
What other positions at this company match my qualifications?	
What are the average starting salaries (benefits)?	
Is traveling or relocation required?	
Where is the business located (home office, other offices)?	
What are the products or services that the employer provides or manufactures?	
What is the mission statement?	

(Continued)

What kind of reputation does this organization have?	
What is the size of the employer's organization relative to the industry?	
What is the growth history of the organization for the past 5, 10, or 15 years?	
How long has the employer been in business?	
Who is the employer's competition?	

Activity 15.2

Using information from Exercise 15.2, write a personal commercial.

_____

_____

_____

_____

_____

Activity 15.3

Write a statement to use during an invitation to an interview that will help you secure all relevant interview information.

_____

_____

Activity 15.4

Using information obtained in your target company research (Activity 15.1), write three common interview questions and answers. Integrate relevant company information and examples in your answers.

Question	Answer
1.	
2.	
3.	

Activity 15.5

Conduct a salary search for a target job. Identify the salary range. Using your research data, write out a statement you could use to negotiate a higher salary.

Lowest Salary	Highest Salary
$	$

Salary Negotiation Statement:

_____

_____

Activity 15.6

Without using an example provided in this chapter, write one question that you can ask at the end of an interview.

_____

_____

16

Career Changes

success • progress • legacy

HOW DO YOU RATE?

Do you know how to properly transition jobs?	True	False
1. It is normal to change jobs within the same company.	☐	☐
2. Being fired is not the same as being laid off.	☐	☐
3. It is appropriate to look for a new job while still employed.	☐	☐
4. With the exception of being fired, when leaving a job, an employee should always write a thank-you note to his or her former boss.	☐	☐
5. Many people start their own business when they are unable to find a job.	☐	☐

▶ If you answered "true" to the majority of these questions, well done. You are aware of important concepts related to career and life changes.

Career Changes

Where do you want to be in five years? While this is a common interview question, it is also part of goal setting. Based on the career and life plan you created in Chapter 1, career changes should be welcome, because they mean you are accomplishing and updating your goals. Career changes are normal and common occurrences. Though most career changes can be controlled, some are unexpected. Make a commitment to become a life-long learner so you have the current knowledge and skills to deal with unexpected change. In this chapter, you will explore the various career changes that may occur and learn how to welcome change as an opportunity for both personal and career growth.

Training and Development

Many companies offer current and new employees **training**, which is learning new skills. The teaching of new skills may be used to promote employees and/or increase their responsibilities. With the increase of technology usage, employee training is important for many companies. Training is usually provided and/or paid for by the company.

Development is enhancing existing skills. In addition to learning new skills through training, make every effort to attend development sessions. Development sessions make employees more diverse in knowledge, skills, and abilities, which provides an advantage when promotional or other opportunities arise in the workplace. Even if you do not think a development session is in your area of expertise, continue expanding your knowledge and skills in other areas. This is especially helpful if you desire a promotion into a management position.

As an employee who is considering a management position, learn not only the skills necessary for the job, but also complementary skills. Also be aware of key duties within other departments. The development of these skills will increase your knowledge and understanding of the company's mission, goals, and overall strategy. When you can see beyond your specific job, you become more aware of how you are contributing to the company's success.

Topic Situation

The marketing department for Destin's company invited all employees to meet in the conference room during lunch hour to learn more about how to conduct a media interview. Destin did not know a lot about marketing and did not think media interviews were a part of his job.

TOPIC RESPONSE
Should Destin attend?
Why or why not?

Continual Learning

In addition to training and development programs offered by a company, there are other ways to improve and increase your skills and knowledge. **Continual learning** is the ongoing process of increasing knowledge in the area of your career and can be accomplished through formal and/or informal learning.

Formal learning is increasing knowledge and skills through traditional venues such as returning to college. This can be accomplished while you continue working. Consider taking one or two night classes a semester while you work. Be cautious about taking too many classes while working full-time because the increased stress may cause you to perform poorly at both work and school. Many colleges offer online classes, which have become increasingly popular for working adults. These classes allow more freedom and flexibility by allowing students to complete coursework on the Internet.

In addition to college, seminars, workshops, and conferences are available, and some provide college credit. Many of these are offered by vendors or industry experts. Although you may have to pay to attend, your company may be willing to reimburse you or share the cost. Seminars, workshops, and conferences are also excellent methods of expanding your professional network.

Informal learning is increasing knowledge through non-traditional education venues, such as reading career-related magazines, newsletters, and electronic articles associated with your job. Another means of informal learning is using the Internet to research career-related information. Informal learning is an ongoing process and can occur during informational interviews, in conversations with professionals in your target career, and by attending association meetings. Make every opportunity a learning opportunity.

Exercise 16.1

What additional courses or career-related information might be helpful to you when you begin working in your target career?

_____

_____

_____

Changes in Employment Status

Throughout this text, we have stressed the importance of goal setting. As you begin meeting your stated career goals, establish new ones. If you are following your life plan, you will have the desire to change jobs as you advance in your area of expertise. Job changes depend on many factors. Some reasons for changing jobs include:

- Acquired experience for an advanced position
- Opportunity for higher salary
- Desire for improved work hours
- Need for increased responsibility, status, and/or power
- Perceived decrease in stress
- Desire for different work environment and/or colleagues

It is normal for employees to move within and outside of a company. A poor economy has forced some employees to change positions and, in some cases, careers. The Bureau of Labor Statistics reports that individuals will have at least six to seven careers in their lifetime, although some will argue that the current economy is driving that number higher. Changes in employment status include promotions, voluntary terminations, involuntary terminations, lateral transfers, and retirement. This section presents and discusses these changes in employment status and provides tips on how to handle each situation in a proactive and professional manner.

New Job Searches

Depending on the work situation, some employees constantly explore new job opportunities, whereas others are forced to find a new job when their current employer eliminates their position. No matter your situation, identify when to share your desire for a new job and when to keep your job search private. If you have recently received a college certification or degree that qualifies you for a higher position, it is appropriate to approach your supervisor or human resource department to inform them of your increased qualifications and desire for additional responsibilities and/or promotion. It is also appropriate to share your need for a new job if a situation is requiring you to move out of the area. In this instance, your employer may have contacts to assist you in securing a new job in another city. If you have had good performance evaluations and are leaving voluntarily, ask your immediate supervisor, another superior, or coworkers if they are willing to serve as references for potential employers. If they agree to serve as references, secure letters of recommendation written on company letterhead. It is helpful to write and provide a draft letter for your reference that highlights your accomplishments and favorable work attitude. Finally, if you have mastered your job duties, have had good performance evaluations, and are beginning to feel bored, respectfully share your desire for increased responsibilities with your boss.

Apart from the previously mentioned circumstances, do not share your desire to change jobs with anyone at work, including close coworkers. Sharing secrets at work can be used against you. Therefore, keep your job search private. Conduct your job search outside of work hours and schedule job interviews before or after work.

Exercise 16.2

List at least four key qualities, skills, or accomplishments you would want included in a draft letter of recommendation from your employer. If possible, provide examples.

_____

_____

_____

Grace and style are two key words to remember if coworkers learn you are looking for a new position either within or outside of your company. When confronted about your job search, be brief and positive. State that you desire a move but keep your explanation simple. You do not have to share details as to why you want to move on. It is also not necessary to share details about potential employers or the status of your job search.

Promotions

A **promotion** is when an employee moves to a position higher in the organization with increased pay and responsibility. The first step in securing a future promotion within your company is to begin behaving and dressing for advancement. Secure a copy of the job description and/or research key skills necessary for your desired position. Begin acquiring work experience in the target area by volunteering for assignments that provide the needed experience. Join committees and assist with projects that will develop and increase your network. Develop new skills through appropriate classes, job training, and other educational experiences to increase your qualifications. Watch and learn from those who are already in the position you desire. Implement this plan and you will gain the necessary qualifications and have the experience when an advanced position becomes available.

If you receive a promotion, congratulate yourself. Your hard work has been noticed by others within your company and they want to reward your excellent behavior. A promotion also means that you are advancing toward your career goals. When you are promoted, thank your former boss either verbally or with a simple handwritten thank-you note. Communicate to your former boss how he or she has helped you acquire new skills. Be sincere. Even if your former boss was less than perfect, his or her behavior taught you how to lead and manage. Keep your note positive and professional. With your promotion, you most likely will see an increase in pay, a new title, and new responsibilities. If your promotion occurred within the same company, do not gloat; there were probably others within the company who also applied for the job. Behave in a positive, pleasant, and professional manner that reinforces that your company made the right choice in selecting you for the position.

In your new job, do not try to reinvent the wheel. Become familiar with the history of your department or area. Be sensitive to the needs and adjustments of your staff. Review files and begin networking with people who can assist you in achieving department goals. When you are new to a position, you do not know everything. Ask for and accept help from others.

TOPIC RESPONSE

Based on everything you have learned in this course, what additional activities can Rachel do to position herself for a promotion?

Topic Situation

With a history of favorable performance evaluations, Rachel wants a promotion. She decided to take responsibility and began evaluating potential positions within her company for which she might qualify. While conducting research, Rachel created a list of additional knowledge, skills, and abilities needed for the promotion.

Voluntary Terminations

Leaving a job on your own is called a **voluntary termination**. Voluntary terminations frequently occur when an employee has taken a job with a new employer or when retiring. Although at times the workplace can be so unbearable that you want to quit without having another job, it is best to not quit unless you have another job waiting. No matter what the situation, when voluntarily leaving a job, be professional and do not burn bridges.

When taking a voluntary termination, resign with a formal letter of resignation. A **letter of resignation** is a written notice of a voluntary termination. Unless you are working with a contract that specifies an end date of your employment, you are technically not required to provide advance notice of your voluntary termination. It is, however, considered unprofessional to resign from work without notice. Typically, two weeks' notice is acceptable. State your last date of employment in your letter of resignation. Include a positive statement about the employer and remember to sign and date your letter. Figure 16.1 is a sample letter of resignation.

In your final days of employment, do not speak or behave negatively. Leave in a manner that would make the company want to rehire you tomorrow. Coworkers may want to share gossip or speak poorly of others, but remain

February 1, 2018

Susie Supervisor
ABC Company
123 Avenue 456
Anycity, USA 98765

Re: Notice of Resignation

Dear Ms. Supervisor:

While I have enjoyed working for ABC Company, I have been offered and have accepted a new position with another firm. Therefore, my last day of employment will be February 23, 2018.

In the past two years, I have had the pleasure of learning new skills and of working with extremely talented individuals. I thank you for the opportunities you have provided me and wish everyone at ABC Company continued success.

Sincerely,

Jennie New-Job

Jennie New-Job
123 North Avenue
Anycity, USA 98765

Figure 16.1

Sample Resignation Letter

professional by refraining from this type of conversation. It is also inappropriate to damage or take property that belongs to the company. Do not behave unethically. Take only personal belongings, and leave your workspace clean and organized for whoever assumes your position. Preserve the confidentiality of your coworkers, department, and customers.

Audrey had a coworker who had been looking for a job over the past few months. Audrey knew this because the coworker not only told everyone, but also used the company equipment to update, print, and mail her résumé. Audrey often heard the coworker talking to potential employers on the telephone. On the day Audrey's coworker finally landed a new job, the coworker proudly announced to everyone in the office that she was "leaving the prison" and that afternoon would be her last day at work. The coworker went on to bad-mouth the company, her boss, and several colleagues. As she was cleaning out her desk, Audrey noticed that the coworker started packing items that did not belong to her. When Audrey shared this observation, the coworker said she deserved the items and that the company would never miss them. A few weeks later, Audrey's former coworker came by the office to say hello. Audrey asked her how her new job was going. "Well . . ." said the coworker, "the job fell through." The coworker explained that she was stopping by the office to see if she could have her old position back. Unfortunately, the former coworker left in such a negative manner that the company would not rehire her.

On the last day of employment with your company, it is common to meet with a representative from the human resource department or with your immediate supervisor to receive your final paycheck. This paycheck should include all unpaid wages and accrued vacation. This is also when you will formally return all company property, including your keys and name badge. You may receive an exit interview. An **exit interview** is when an employer meets with an employee who is voluntarily leaving a company to identify opportunities for improving the work environment. During this interview, a company representative will ask questions regarding the job you are leaving, the boss, and the work environment. The company's goal is to secure any information that provides constructive input on how to improve the company. Share opportunities for improvement, but do not turn your comments into personal attacks. Although it is sometimes tempting to provide negative information in an exit interview, remain positive and professional.

Involuntary Terminations

An **involuntary termination** is when an employee loses his or her job. Types of involuntary terminations include firings and layoffs. A **firing** occurs when an employee is terminated due to a performance issue. A **layoff** is when a company terminates employment through no fault of the employee. Commonly, layoffs are a result of a company's financial inability to keep a position, so the position is eliminated. Some layoffs are a result of **restructuring**, which is when the company has eliminated a position due to a change in strategy.

If you are fired, you have lost your job as a result of a performance issue. Unless you have done something outrageous (such as blatant theft or harassment), you should have received a poor performance warning prior to being fired. Typically, this progressive discipline includes a verbal and/or written warning prior to termination. If you are totally unaware of why you are being fired, ask for documentation to support the company's decision. Firing based on outrageous behavior will be supported by a policy. Performance issues should be supported by prior documentation. When you are informed of your firing, you should immediately receive your final paycheck. You will also be asked to immediately return all company property, including keys and your name badge. Although you may be angry or caught off guard, do not make threats against the company or its employees and do not damage company property. Doing so is not only immature, but punishable by law. Remain calm and professional. If you think you are being wrongfully terminated, your legal recourse is to seek assistance from your state's labor commission or a private attorney.

Many people consider a layoff a form of firing. This is not true. Firing is a result of poor performance. A layoff is a result of a company's change of strategy or its inability to financially support a position. Although some companies lay off employees based on performance, most do it on seniority. Frequently, when the company's financial situation improves, employees may be recalled. A **work recall** is when employees are called back to work after being laid off. If you have been laid off, remain positive and ask your employer for a letter of reference and job search assistance. This job search assistance may include help with updating a résumé, counseling, job training, and job leads. Some companies require employees to take unpaid work days, called **furloughs**. Work furloughs are not a result of poor performance. They are a result of employers trying to save financial resources. If your company implements a work furlough program, make the best of the situation. Be happy you still have a job and find ways to assist the company in improving its financial situation. Knowing your current employer is experiencing economic challenges provides you an opportunity to update your résumé and create a plan should your employer need to take additional steps toward saving resources.

In today's competitive environment, it is common for companies to restructure. As stated earlier, restructuring involves a company changing its strategy and reorganizing resources. This commonly results in eliminating unnecessary positions. If your position is eliminated, remain positive and inquire about new positions. In a restructuring situation, it is often common for new positions to be created. Once again, do not bad-mouth anyone or openly express your anger or dissatisfaction over the situation. If you have recently acquired new skills, now is the time to communicate and demonstrate them. Keep a record of your workplace accomplishments and keep your ears open for new positions within the company for which you qualify.

> **Think About It**
>
> What is the best way to use your time during a furlough day?

Other Moves within an Organization

In addition to promotions and terminations, there are several other methods of moving within and outside the company. These include lateral moves, demotions, and retirement. A **lateral move** is when an employee is transferred to another area of an organization with the same level of responsibility. Lateral moves involve only a change in department or work area. A change in pay is not involved in a lateral move. If you are moved to a different position and experience a pay increase, it is considered a promotion. A **demotion** is when an employee is moved to a position with less responsibility and experiences a pay decrease. Although demotions are rare, they can occur when an employee's performance is not acceptable and the employee chooses to not leave the company. Demotions are not always a result of poor performance; they sometimes occur to avoid a layoff. Of all the changes an employee can make, a demotion is by far the most difficult. You experience not only a decrease in pay, but also a decrease in job title and status. If you are demoted, remain professional and be respectful of your new boss. Today's uncertain economic environment has caused a shift in the traditional career ladder. There may be a time where you may request a lateral move or demotion to learn a new skill or industry. In this case, the individual is taking a demotion to prepare for a future promotion. Though risky, these non-traditional career moves are becoming more common.

The final change in employment status is called **retirement**. Retirement is when an employee is voluntarily leaving employment and will no longer be working. Although this text addresses those entering the workforce, it is never too early to start planning for your retirement, both mentally and financially. This can be done by establishing career goals and deadlines, in addition to contributing to a retirement fund.

Entrepreneurship

Some individuals do not want to work for others and have a desire to be their own boss. A final and common form of career transition is that of becoming an entrepreneur. An **entrepreneur** is someone who assumes the risk of succeeding or failing by owning and operating a business. Although owning and operating your own business may sound glamorous, doing so involves work. Individuals become entrepreneurs for several reasons. The most common reason is when someone has identified a business opportunity he or she wants to exploit. People also become entrepreneurs because they would rather work for themselves, want more control of their work environment, want more income, or lost their jobs and have been unable to find another.

Individuals with full-time jobs sometimes supplement their income either by running a business on the side or by taking on a second job. Although this is a good way to earn extra income, do not run a side business or hold a second job that competes with or utilizes your employer's resources or confidential information. Check with your HR department to ensure outside employment is allowed. If your current employer permits employees to hold a second job or run a side business, do not allow this activity to interfere with your full-time employment. Keep the ventures separate.

Entrepreneurship is a rewarding career option for many and plays a valuable role in the U.S. economy. There are various methods of becoming an entrepreneur. You can start your own business, purchase an existing business, or operate a franchise. From a home-based business to running a chain of retail sites, every entrepreneur started with a dream. To be a successful entrepreneur, you need to have a passion for your business. You also need to know how to plan, manage finances, and make yourself creatively and professionally stand out from a crowd. These are all skills you have started to develop by reading this text.

Web Search

Research entrepreneurship online to see if entrepreneurship is for you.

Talk It Out

What type of business would you like to own? What steps do you need to take to make this dream a reality?

If you are interested in becoming an entrepreneur, there are many resources available to you. Start by exploring the Small Business Administration government website. Here you will find online, local, and national resources to get you started on the road to entrepreneurial success.

Career Success

As you have learned, there are several methods of advancing your career both within and outside of an organization. Although it is not good to change jobs too frequently, those with healthy careers rarely stay in one position their entire career. Personal issues frequently influence the choices we make in our careers. Health matters, changes in marital status, children, and elder care are just a few of these issues. There is a French proverb that states that some work to live, but others live to work. Although there are trade-offs, your personal life must be a priority. Any change in your career will not only affect you, but those to whom you are close. Therefore, make them a consideration in your career decisions.

As you progress in your career, you will experience a growth in income and, hopefully, a greater appreciation and understanding of the world around you. Take time to review and update your life plan and consider allocating a portion of your time, talent, and/or financial resources to causes that benefit others. Consider a favorite non-profit organization or situation that can benefit from your assistance, and then make a commitment to make a positive impact in someone else's life. Ensure this commitment is followed through by including it in your life plan and personal budget.

Career success is all about personal choice and maintaining an attitude of success. Regardless of when you plan to advance your career, keep your résumé updated, make a commitment to continuous learning, and display leadership. Doing so keeps you motivated to take on additional responsibilities and increases your knowledge, skills, and abilities. This will prepare you if some unforeseen opportunity comes your way. Keep focused on your life plan and consistently display professionalism and you will be positioned for a lifetime of workplace success.

MyStudentSuccessLab Please visit **MyStudentSuccessLab**: Anderson|Bolt, Professionalism Skills for Workplace Success, 4/e for additional activities, resources, and outcomes assessments.

Workplace Dos and Don'ts

Do continually update your skills and knowledge through training and development	*Don't* assume additional skills and knowledge are not necessary for advancement
Do keep an open mind for job advancement opportunities	*Don't* openly share your dissatisfaction for your current job

Do write a formal resignation letter when leaving a company and a thank-you letter to a boss or mentor when receiving a promotion	*Don't* leave your job without providing adequate notice to your current employer
Do behave professionally when leaving a position	*Don't* take or ruin company property when leaving a position
Do provide valuable feedback and opportunities for improvement during an exit interview	*Don't* turn an exit interview into a personal attack on your former boss or coworkers

Concept Review and Application

You are a Success ful Student if you:

- Identify the importance of continual learning and its role in your career
- Research and identify an entrepreneurial opportunity that interests you and the initial steps to take to make it happen
- Name three specific areas of professionalism you wish to improve and create a plan to do so

Summary of Key Concepts

- Continue learning new skills to help reach your career potential.
- Formal learning is another way to increase skills and knowledge.
- Changes in employment status include promotions, voluntary terminations, involuntary terminations, lateral moves, and retirement.
- Be cautious about sharing your desire for a new job with coworkers.
- There are two types of terminations: voluntary and involuntary.
- When leaving voluntarily, submit a letter of resignation.
- When leaving in an involuntary manner, do not burn bridges or behave in an unprofessional or unethical manner.
- There is a difference between being fired and being laid off.
- It is never too early to begin planning for your retirement.
- Becoming an entrepreneur is an additional form of career transition.

Self-Quiz MATCHING KEY TERMS

Match the key term to the definition using the identifying number.

Key Terms	Answer	Definitions
Continual learning		1. When an employee is moved to a position with less responsibility and experiences a pay decrease
Demotion		2. Unpaid work days that employees are required to take
Development		3. Written notice of a voluntary termination
Entrepreneur		4. When an employee is terminated due to a performance issue
Exit interview		5. When an employee loses his or her job
Firing		6. Someone who assumes the risk of succeeding or failing in business
Formal learning		7. When employees are called back to work after being laid off
Furlough		8. Leaving a job on your own
Informal learning		9. Increasing knowledge through traditional venues
Involuntary termination		10. Learning new skills
Lateral move		11. When an employee is voluntarily leaving employment and will no longer be working
Layoff		12. When an employer interviews an employee who is voluntarily leaving
Letter of resignation		13. When a company terminates employment through no fault of the employee
Promotion		14. Enhancing existing skills
Restructuring		15. Increasing knowledge through non-traditional education venues
Retirement		16. When a company has eliminated a position due to a change in strategy
Training		17. When an employee moves to a position higher in the organization with increased pay and responsibility
Voluntary termination		18. When an employee is transferred to another area of an organization with the same level of responsibility
Work recall		19. Ongoing process of increasing knowledge

Think Like a Boss

1. Why would it be important to encourage training and development sessions within your department?

2. You hear through the grapevine that one of your best employees is looking for another job. What should you do?

3. Management has told you that you must lay off four of your employees. How do you determine whom to lay off and how best to tell them? How do you defend your decision?

Activities

Activity 16.1

Based on the career plan you created in Chapter 1, identify additional training, development, and continual learning you will need for professional success.

Training	Development	Continual Learning

Activity 16.2

Throughout this text, you have learned the importance of maintaining positive human relations in the workplace. Name at least three activities to implement that will decrease your chances of being laid off if layoffs were necessary within your company.

_____

_____

_____

Activity 16.3

Write a draft letter of recommendation for yourself.

_____

_____

_____

_____

_____

_____

_____

Activity 16.4

Based on your target career, research and identify two professional conferences you would like to attend and two professional associations you would like to join.

Conferences

1. _____

2. _____

Associations

1. _____

2. _____

Activity 16.5

Identify an issue in your life that bothers you or you see as a problem others face. How can you turn this problem into an entrepreneurial venture?

_____

_____

_____

Activity 16.6

Reflecting on the primary topics of self-management, workplace basics, relationships, and career planning tools, where are your weaknesses in each area and what specific skills can you develop in each area listed below to increase your value as an employee?

	Weakness	Increased Skill
Self-management		
Workplace basics		
Relationships		
Career planning tools		

accommodating conflict management style: a conflict management style that allows the other party to have his or her own way without knowing there was a conflict

accountability: report back to whoever empowered you to carry out a specific responsibility

active listening: when a receiver provides a sender full attention without distraction

adjourning stage: when team members bring closure to a project

advanced skill set résumé layout: a résumé layout used by those with extensive career experience that emphasizes related work experience, skills, and significant accomplishments

aggressive behavior: the behavior of an individual who stands up for his or her rights in a manner that violates others' rights in an offensive manner

assertive behavior: the behavior of an individual who stands up for his or her rights without violating the rights of others

assets: items that you own that are worth money

attitude: a strong belief about people, things, and situations

autocratic leaders: leaders who make decisions on their own without input from others

automatic deduction plan: when funds are automatically deducted from an employee's paycheck and placed into a bank account

avoiding conflict management style: a passive conflict management style used when one does not want to deal with the conflict so the offense is ignored

bargaining agreement: a contract between an employer and a union that addresses salaries, benefits, working conditions and other common employee matters

behavioral interview question: interview question that asks candidates to share a past experience related to a specific workplace situation

board of directors: a group of individuals responsible for developing the company's overall strategy and major policies

brainstorming: a problem-solving method that involves identifying alternatives that allow members to freely add ideas while other members withhold comments on the alternatives

budget: a detailed financial plan used to allocate money for a specific time period

business letter: a formal written form of communication used when a message is being sent to an individual outside of an organization

business memo: written communication sent within an organization (also called *interoffice memorandum*)

capital budget: a financial plan used for long-term investments including land and large pieces of equipment

career objective: an introductory written statement used on a résumé for individuals with little or no work experience

casual workdays: workdays when companies relax the dress code policy

chain of command: identifies who reports to whom within the company

character: the unique qualities of an individual, which usually reflect personal morals and values

charismatic power: a type of personal power that makes people attracted to you

coercive power: power that uses threats and punishment

collaborating conflict management style: a conflict management style in which both parties work together to arrive at a solution without having to give up something of value

communication: the process of a sender transmitting a message to an individual (receiver) with the purpose of creating mutual understanding

competent: having the ability to answer questions when a customer asks

compromising conflict management style: a conflict management style that is used when both parties give up something of importance to arrive at a mutually agreeable solution to the conflict

conference calls: meetings utilizing technology and are treated like face-to-face prearranged meetings

confidential: matters that should be kept private

conflict: a disagreement or tension between two or more parties (individuals or groups)

conflict of interest: when someone influences a decision that directly or indirectly benefits him or her

connection power: based on using someone else's legitimate power

continual learning: the ongoing process of increasing knowledge in the area of your career

corporate culture (organizational culture): values, expectations, and behavior of people at work; the company's personality being reflected through employees' behavior

cost of living: average cost of basic necessities such as housing, food, and clothing for a specific geographic area

courtesy: exercising manners, consideration, and respect toward others

cover letter: a letter that introduces your résumé

creativity: the ability to produce something new and unique

credit report: a detailed credit history on an individual

culture: different behavior patterns of various groups

customer: an individual or business that buys a company's product (good or service)

customer service: the treatment an employee provides the customer

debt: money owed

decoding: how a receiver interprets a message

delegate: when a manager or leader assigns part or all of a project to someone else

democratic leaders: leaders who make decisions based upon input from others

demotion: when an employee is moved to a lower position with less responsibility and a decrease in pay

dental benefits: insurance coverage for teeth

department: sub-area of a division that carries out specific tasks respective of its division

dependable: being reliable and taking responsibility to assist a customer

development: sessions to enhance or increase existing skills

direct benefits: monetary employee benefits

directional statements: a company's mission, vision, and values statements; these statements are the foundation of a strategic plan explaining why a company exists and how it will operate

diversity statements: corporate statements that remind employees that diversity in the workplace is an asset and not a form of prejudice and stereotyping

division: how companies arrange major business functions

documentation: a formal record of events or activities

dress code: an organization's policy regarding appropriate workplace attire

electronic job search portfolio (e-portfolio) is a computerized folder that contains electronic copies of all job search documents

employee assistance program (EAP): an employee benefit that typically provides free and confidential psychological, financial, and legal advice

employee handbook: a formal document provided by the company that outlines an employee's agreement with the employer regarding work conditions, policies, and benefits

employee loyalty: an employee's obligation to consistently support a company and its mission

employee morale: the attitude employees have toward the company

employee orientation: a time when a company provides new employees important information including the company's purpose, structure, major policies, procedures, benefits, and other matters

employment-at-will: policies that do not contractually obligate employees to work for the company for a specified period

empowerment: pushing power and decision making to the individuals who are closest to the customer in an effort to increase quality, customer satisfaction, and, ultimately, profits

encoding: identifying how a message will be sent (verbally, written, or nonverbally)

entrepreneur: someone who assumes the risk of succeeding or failing in business through owning and operating the business

ethics: a moral standard of right and wrong

ethics statement: a formal corporate policy that addresses the issue of ethical behavior and punishment should someone behave inappropriately

ethnocentric: when an individual believes his or her culture is superior to others

etiquette: a standard of social behavior

executive presence: having the attitude of an executive by demonstrating appropriate workplace behavior

executives (senior managers): typically have title of vice president; individuals who work with the president of a company in identifying and implementing the company strategy

exit interview: when an employer meets with an employee who is voluntarily leaving a company to identify opportunities to improve the work environment

expense: money going out

expert power: power that is earned by one's knowledge, experience, or expertise

extrinsic rewards: rewards that come from external sources including such things as money and praise

Fair Isaac Corporation (FICO) score: the most common credit rating

feedback: when a receiver responds to a sender's message based upon the receiver's interpretation of the original message

finance and accounting: a function that is responsible for the securing, distribution, and growth of the company's financial assets

firing: when an employee is terminated because of a performance issue

fixed expenses: expenses that do not change from month to month

flexible expenses: expenses that change from month to month

forcing conflict management style: a conflict management style that attempts to make the other party do things your way, the other party has no say whatsoever

formal communication: workplace communication that occurs through memos, meetings, or lines of authority

formal learning: increasing knowledge through traditional venues such as returning to college to increase knowledge, improve skills, or receive an additional or advanced degree

formal teams: developed within the formal organizational structure and may include functional teams or cross-functional teams

forming stage: when team members first get to know each other and form initial opinions about other members

free-reign leaders: leaders who allow team members to make their own decisions without input from the leader (also known as *laissez-faire leaders*)

full-time employee: work a pre-determined number of hours per week and are eligible for employer benefits

functional résumé layout: a résumé layout that emphasizes relevant skills when related work experience is lacking

furloughs: unpaid work days that employees are required to take

glass ceiling: invisible barrier that frequently makes executive positions off-limits to females and minorities, thus prohibiting them from advancing up the corporate ladder through promotions

glass wall: invisible barrier that frequently makes certain work areas such as a golf course off-limits to females and minorities, thus prohibiting them from advancing up the corporate ladder through promotions

goal: a target

good: a tangible product produced by a company

gossip: personal information about another individual that is hurtful and inappropriate

grapevine: an informal communication network where employees talk about workplace issues of importance

grievance: a problem, unfair treatment, or conflict related to employment

grievance procedure: formal steps taken in resolving a conflict between the union and an employer

gross income: the amount of money in a paycheck before paying taxes or other deductions

group: two or more people who share a common interest and have one leader

group interview: an interview that involves several applicants interviewing at the same time while being observed by company representatives

harassment: offensive, humiliating, or intimidating behavior

hostile behavior harassment: any behavior of a sexual nature by another employee that someone finds offensive, including verbal slurs, physical contact, offensive photos, jokes, or any other offensive behavior of a sexual nature

human relations: interactions occurring with and through people

human resource management (HRM): a business function that deals with recruiting, hiring, training, evaluating, compensating, promoting, and terminating employees

implied confidentiality: an obligation to not share information with individuals with whom the business is of no concern

income: money coming in

incompetent boss: a boss who does not know how to do his or her job

indirect benefits: nonmonetary employee benefits such as health care and paid vacations

informal communication: workplace communication that occurs among individuals without regard to the formal lines of authority

informal learning: increasing knowledge through non-traditional education venues such as reading career-related magazines, newsletters, and other articles associated with a job

informal team: group of individuals who get together outside of the formal organizational structure to accomplish a goal

information heading: a résumé heading that contains relevant contact information including name, mailing address, city, state, ZIP code, contact phone, and e-mail address

information power: power based upon an individual's ability to obtain and share information

information systems (IS): a business function that deals with the electronic management of computer-based information within the organization

informational interview: when a job seeker meets with a business professional to learn about a specific career, company, or industry

innovation: the process of turning a creative idea into reality

integrity: when someone consistently behaves in an ethical manner

interest: the cost of borrowing money

interview portfolio: a folder to be taken on an interview that contains photocopies of documents and items pertinent to a position

intrinsic rewards: internal rewards that include such things as self-satisfaction and pride of accomplishment

introductory period: (also known as probation and orientation period), is typically the first one to three months of employment where employers are provided time to evaluate a new hire's performance and determine if the new hire should continue as an employee

involuntary termination: when an employee loses his or her job

job burnout: a form of extreme stress where you lack motivation and no longer have the desire to work

job description: a document that outlines specific job duties and responsibilities for a specific position

job search portfolio: a collection of paperwork needed for job searches and interview

job-specific skills: skills that are directly related to a specific job or industry

labeling: when one describes an individual or group of individuals based upon past actions

laissez-faire leaders: leaders who allow team members to make their own decisions without input from the leader (also known as *free reign leaders*)

lateral move: when an employee is transferred to another area (department) of an organization with the same level of responsibility

layoff: when a company terminates employment through no fault of the employee

leadership: a process of one person guiding one or more individuals toward a specific goal

learning style: the method of how you best take in information and/or learn new ideas

legal counsel: a function within a business that handles all legal matters relating to the company

legitimate power: the power that is given to an employee from the company

letter of recommendation: a written testimony from another person that states that a job candidate is credible

letter of resignation: a written notice of your voluntary termination

letterhead: quality paper that has the company logo, mailing address, and telephone numbers imprinted on it

levels of ethical decisions: three questions to help you make an ethical decision

liability: an obligation to pay what you owe

life plan: a written document that identifies goals in all areas of your life

listening: the act of hearing attentively

loan: a large debt that is paid in smaller amounts over a period of time and has interest added to the payment

locus of control: identifies who you believe controls your future

long-term goal: a target that takes longer than one year to accomplish

marketing: a business function responsible for creating, pricing, selling, distributing, and promoting the company's product

Maslow's Hierarchy of Needs: states that throughout one's lifetime, an individual's needs are met as they progress up a pyramid (hierarchy) of five needs

McClelland's Theory of Needs: holds that people are primarily motivated by one of three factors: achievement, power, or affiliation

mediator: a neutral third party whose objective is to assist two conflicting parties in coming to a mutually agreeable solution

medical benefits: insurance coverage for physician and hospital visits

meeting agenda: an outline of all topics and activities that are to be addressed during a meeting

meeting chair: the individual who is in charge of a meeting and has prepared the agenda

mentor: someone who can help an employee learn more about his or her present position, provide support, and help develop the employee's career

middle manager: typically has the title of *director* or *manager;* these individuals work on tactical issues

mirror words: words that describe the foundation of how you view yourself, how you view others, and how you will most likely perform in the workplace

mission statement: a company's statement of purpose

money wasters: small expenditures that consume a larger portion of one's income than expected

morals: a personal standard of right and wrong

motivation: an internal drive that causes people to behave a certain way to meet a need

negative stress: an unproductive stress that affects your mental and/or physical health including becoming emotional or illogical or losing your temper

negotiation: create a dialog with other involved parties in an effort to create a solution that is fair to all involved parties

net income: the amount of money you have after all taxes and deductions are paid

net worth: the amount of money that is yours after paying off debt

network list: an easily accessible list of all professional network contacts' names, industries, addresses, and phone numbers

networking: meeting and developing relationships with individuals outside one's immediate work area; the act of creating professional relationships

noise: anything that interrupts or interferes with the communication process

non-listening: When a receiver fails to make any effort to hear or understand the sender's message

nonverbal communication: what is communicated through body language

norming stage: when team members accept other members for who they are

objectives: short-term goals that are measurable and have specific time lines that occur within one year

one-on-one interview: an interview that involves a one-on-one meeting between the applicant and a company representative

online identity: the image formed when someone is communicating and/or researching you through electronic means such as personal web pages and search engines

open-door policy: communicates to employees that management and the human resource department is available to listen should the employee need to discuss a workplace concern

operational budget: a financial plan used for short-term items including payroll and the day-to-day costs associated with running a business

operational issue: organizational issues that typically occur on a daily basis and/or no longer than one year

operations: a business function that deals with the production and distribution of a company's product

operations manager: first-line manager who is typically called a *supervisor* or *assistant manager*

organizational chart: a graphic visual display of how a company organizes its resources; identifies key functions within the company and shows the formal lines of authority for employees

organizational structure: the way a company is organized

panel interview: an interview that involves the applicant meeting with several company employees at the same time

part-time employee: have varied work hours and normally do not qualify for employer benefits

passive behavior: the behavior exhibited when an individual does not stand up for his or her rights by consistently allowing others to have their way

passive listening: the receiver is selectively hearing parts of a message

perception: one's understanding or interpretation of reality

performance evaluation: a formal appraisal that measures an employee's work performance

performing stage: when team members begin working on their task

personal brand: reflects trains you want others to think of when they think of you

personal commercial: a brief career biography that conveys one's career choice, knowledge, skills, strengths, abilities, and experiences

personal financial management: the process of controlling personal income and expenses

personal profile: an introductory written statement used on a résumé for individuals with professional experience related to their target career

personality: a stable set of traits that assist in explaining and predicting an individual's behavior

politics: obtaining and utilizing power

positive self-talk: a mental form of positive self-reinforcement that helps remind you that you are qualified and deserve both the interview and the job

positive stress: productive stress that provides strength to accomplish a task

power: one's ability to influence another's behavior

power words: action verbs that describe your accomplishments in a lively and specific way

prejudice: a favorable or unfavorable judgment or opinion toward an individual or group based on one's perception (or understanding) of a group, individual, or situation

president or chief executive officer (CEO): the individual responsible for operating the company; this individual takes his or her direction from the board of directors

priorities: determine what needs to be done and in what order

procrastination: putting off tasks until a later time

product: what is produced by a company

productivity: to perform a function that adds value to a company

professional network: a group of relationships that are established primarily for business purposes. A professional network is created through networking.

professionalism: workplace behaviors that result in positive business relationships

profit: revenue (money coming in from sales) minus expenses (the costs involved in running the business)

projection: the way you feel about yourself is reflected in how you treat others

promotion: moving to a position higher in the organization with increased pay and responsibility

protected class: a group of individuals who are protected from discrimination based upon civil rights legislation.

proxemics: the study of distance (space) between individuals

quality: a predetermined standard that defines how a good is to be produced or a service is to be provided

quid pro quo harassment: a form of sexual harassing behavior that is construed as reciprocity or payback for a sexual favor

race: a group of individuals with certain physical traits

reciprocity: creating debts and obligations for doing something

respect: holding someone in high regard

responsibility: accepting the power that is being given and the obligation to perform

responsive: being aware of a customer's needs, often before the customer

restructuring: when a company eliminates a position due to a change in corporate strategy

résumé: a formal written profile that presents a person's knowledge, skills, and abilities to potential employers

retirement: when an employee voluntarily leaves the company and will no longer work

retirement plan: a savings plan for when you retire and permanently leave the workforce

reward power: the ability to influence someone with something of value

right to revise: a statement contained in an employee handbook that provides an employer the opportunity to change or revise existing policies

Robert's Rules of Order: a guide to running meetings, oftentimes referred to as *parliamentary procedure*

self-concept: how you view yourself

self-discovery: the process of identifying key interests and skills built upon career goals

self-efficacy: your belief in your ability to perform a task

self-esteem: how confident one is in his/her self and his/her abilities

self-image: your belief of how others view you

sender: an individual wanting to convey a message

senior managers or executives: individuals who work with the president in identifying and implementing the company strategy

service: an intangible product produced by a company that one cannot touch or see

sexual harassment: unwanted advances of a sexual nature

shop steward: a coworker who assists others with union-related issues and procedures

short-term goals: goals that can be reached within a year's time (also called *objectives*)

slang: an informal language used among a particular group

SMART goal: a goal that is specific, measurable, achievable, relevant, and time-based

soft skills: people skills that are necessary when working with others in the workplace

stereotyping: making a generalized image of a particular group or situation

storming stage: when team members have conflict with each other

strategic issues: major company goals that typically range from three to five years or more

strategic plan: a formal document that is developed by senior management; the strategic plan identifies how the company secures, organizes, utilizes, and monitors its resources

strategy: a company's road map for success that outlines major goals and objectives

stress: a body's reaction to tense situations

structured interview question: a type of interview question that addresses job-related issues where each applicant is asked the same question

supervisor: first-line manager who concerns him- or herself with operational issues

synergy: two or more individuals working together and producing more than the sum of their individual efforts

tactical issues: business issues that identify how to link the corporate strategy into the reality of day-to-day operations; the time line for tactical issues is one to three years

targeted job search: job search process of discovering positions for which you are qualified in addition to identifying specific companies for which you would like to work

team: a group of people linked to a common purpose

technology-use policy: an outline of expectations including privacy, liability, and potential misconduct issues related to the use of company technology.

teleconference: an interactive communication that connects participants through the telephone without the opportunity of visually seeing all participants

temporary employee: an employee who is hired only for a specified period of time, typically to assist with busy work periods or to temporarily replace an employee on leave

time management: how you manage your time

trade-off: giving up one thing to do something else

training: learning new skills

transferable skills: skills that can be transferred from one job to another

union: a third-party organization that protects the rights of employees and represents employee interests to an employer

union contract: the formal document that addresses specific employment issues including the handling of grievances, holidays, vacations, and other issues

unstructured interview question: a probing, open-ended interview question intended to identify if the candidate can appropriately sell his or her skills

value: getting a good deal for the price paid for a product

values: things that are important to an individual

values statement: part of a company strategic plan that defines what is important to (or what the priorities are for) the company

verbal communication: the process of using words to send a message

video conference: a form of interactive communication using two-way video and audio technology

virtual teams: teams that function through electronic communications because they are geographically dispersed

vision benefits: insurance coverage for vision (eye) care

vision statement: part of a company's strategic plan that describes the company's viable view of the future

voluntary termination: leaving a job on your own

Vroom's Expectancy Theory: holds that individuals will behave in a certain manner based upon the expected outcome

work recall: when employees are called back to work after a layoff

work wardrobe: clothes that are primarily worn to work and work-related functions

workplace bullies: employees who are intentionally rude and unprofessional to coworkers

workplace discrimination: acting negatively toward someone based on race, age, gender, religion, disability, or other areas

workplace diversity: differences among coworkers including culture, race, age, gender, economic status, and religion

written communication: a form of business communication that is printed, handwritten, or sent electronically

Suggested Readings

Chapter 1

Rotter, J. B. "Generalized Expectancies for Internal versus External Control of Reinforcement." *Psychological Monographs*, Vol. 80, No. 1 (1966): 1–28.

Taylor, M. "Does Locus of Control Predict Young Adult Conflict Strategies with Superiors? An Examination of Control Orientation and the Organizational Communication Conflict Instrument." *North American Journal of Psychology*, Vol. 12, No. 3 (2010): 445–458.

Bandura, A. (1994). Self-efficacy. In V. S. Ramachaudran (Ed.), *Encyclopedia of Human Behavior*, Vol. 4. New York: Academic Press, pp. 71–81.

Bandura, A. "Human Agency in Social Cognitive Theory." *American Psychologist*, Vol. 44, No. 9 (1989): 1175–1184.

"Work-Family Conflicts Affect Employees at All Income Levels," *HR Focus* 87 (April 2010): 9.

Golden, E. Organizational Renewal Associates. 1971. Golden LLC, May 2011, www.goldenllc.com

http://www.forbes.com/sites/actiontrumpseverything/2013/01/13/how-to-plan-your-life-when-you-cant-plan-your-life/

Hoel, H., Glasco, L., Hetland, J., Cooper, C. L., and Einarsen, S. "Leadership Styles as Predictors of Self-reported and Observed Workplace Bullying." *British Journal of Management*, Vol. 21, No. 2 (2010): 453–468.

Newhouse, N. "Implications of Attitude and Behavior Research for Environmental Conservation." *Journal of Environmental Education*, Vol. 22, No. 1 (Fall 1990): 26–32.

Savickas, Mark L. "Career Studies as Self-Making and Life Designing." *Career Research and Development* 23 (2010): 15-18.

Robins, E.M. (2014, April 24). An Instructional Approach to Writing SMART Goals. PowerPoint presented at the 19th Annual Technology, Colleges, and Community Worldwide Online Conference.

Platt, G. "SMART Objectives: What They Mean and How to Set Them." *Training Journal* (August 2002): 23.

Chapter 2

https://bigfuture.collegeboard.org/pay-for-college/financial-aid-101/financial-aid-faqs. Retrieved May 24, 2014.

Fair Isaac Corporation, Minneapolis, MN, www.myfico.com

Federal Reserve Bank, Washington, DC, www.federalreserve.gov

Annual Credit Report Request Service, Atlanta, GA, www.annualreditreport.com

Kim, J., Sorhaindo, B., and Garman, E. T. Relationship between Financial Stress and Workplace Absenteeism of Credit Counseling Clients. *Journal of Family and Economic Issues*, Vol. 27, No. 3 (2006): 458–478, doi:10.1007/s10834-006-9024-9

Peete, S. S. "Employers Are Stung with a Hefty Price When Employees Suffer an Identity Theft," *Supervision*, Vol. 69, No. 7 (July 2008): 10–12.

The Motley Fool, Alexandria, VA, www.fool.com (how to invest, personal finance)

The Lampo Group, Inc. Brentwood, TN, www.daveramsey.com

Collins, J. M. and O'Rourke, C. M. (2010), Financial Education and Counseling—Still Holding Promise. *Journal of Consumer Affairs*, 44: 483–498. doi: 10.1111/j.1745-6606.2010.01179.x

Bernerth, Jeremy B.; Taylor, Shannon G.; Walker, H. Jack; Whitman, Daniel S. An empirical investigation of dispositional antecedents and performance-related outcomes of credit scores. *Journal of Applied Psychology*, Vol 97(2), Mar 2012, 469-478. doi: 10.1037/a0026055

Justine S. Hastings & Brigitte C. Madrian & William L. Skimmyhorn, 2013. "Financial Literacy, Financial Education, and Economic Outcomes," *Annual Review of Economics*, Annual Reviews, vol. 5(1), pages 347-373, 05.

Cavanaugh, Afton R. "Rich Dad vs. Poor Dad: Why Leaving Financial Education to Parents Breeds Financial Inequality & Economic Instability." (2013).

Beck, Ted. "Offering financial education in the workplace benefits both employees and employers." *Employment Relations Today* 37.2 (2010): 9-14.

Lusardi, A., Mitchell, O. S. and Curto, V. (2010), Financial Literacy among the Young. *Journal of Consumer Affairs*, 44: 358–380. doi: 10.1111/j.1745-6606.2010.01173.x

Federal Student Aid, An Office of the U.S. Department of Education, "What are the differences between federal and private student loans?" http://studentaid.ed.gov/types/loans/federal-vs-private

Chapter 3

http://www.stress.org/workplace-stress/

http://www.forbes.com/sites/work-in-progress/2012/08/02/stress-at-work-is-bunk-for-business/

http://www.cdc.gov/features/dssleep/

Glynn, Anthony. *Is absenteeism related to perceived stress, burnout levels and job satisfaction?*. Diss. Dublin Business School, 2014.

Page, Matthew, and Matthew Page. *Agent-Based Modelling of Stress and Productivity Performance in the Workplace*. Diss. 2013.

Rhoade, Collin, and Shawn M. Carraher. "Strategic Knowledge Worker Productivity and Leisure Time." *Allied Academies International Internet Conference*. Vol. 14. 2012.

Gallagher, Vickie Coleman. "Managing Resources and Need for Cognition: Impact on Depressed Mood at Work." *Personality and Individual Differences* 53.4 (2012): 534-537.

Cole, Michael S., et al. "Job Burnout and Employee Engagement A Meta-Analytic Examination of Construct Proliferation." *Journal of management* 38.5 (2012): 1550-1581.

Rosekind, M. R., and Gregory, K. B., et. al. "The Cost of Poor Sleep: Workplace Productivity Loss and Associated Costs." *Journal of Occupational & Environmental Medicine*. Vol. 52, No. 1 (2010): 91–98, doi: 10.1097/JOM.0b013e3181c78c30

Finkelstein, E.A., et al. "The Costs of Obesity in the Workplace." *Journal of Occupational and Environmental Medicine* 52.10 (2010): 971-976.

Leeds, R. One Year to an Organized Work Life. (Cambridge, MA: Da Capo Press, 2008).

Knight, C. and Haslam, S.A. "The Relative Merits of Lean, Enriched, and Empowered Offices: An Experimental Examination of the Impact of Workspace Management Strategies on Well-Being and Productivity." *Journal of Experimental Psychology: Applied* 16.2 (2010): 158.

DeVries, G. "Innovations in Workplace Wellness: Six New Tools to Enhance Programs and Maximize Employee Health and Productivity." *Compensation Benefits Review*, Vol. 42, No. 1 (January/February 2010): 46–51.

http://www.smartceo.com/3-ways-entrepreneurs-can-reduce-stress/ (March, 2013)

Centers for Disease Control and Prevention/National Institute for Occupational Safety and Health, www.cdc.gov/niosh/

Forbes, "Stress at Work is Bunk for Business" by Judy Martin, http://www.forbes.com/sites/work-in-progress/2012/08/02/stress-at-work-is-bunk-for-business

Mayo Clinic, exercise and stress: "Get moving to manage stress," http://www.mayoclinic.com/health/exercise-and-stress/SR00036

Chapter 4

Sarantakis, M. (2013, July 11). Why do millennials seem to lack manners? Retrieved from http://www.examinier.com

The Emily Post Institute, Burlington, VT, www.emilypost.com

Velasco, M.S. "More than just good grades: candidates' perceptions about the skills and attributes employers seek in new graduates." *Journal of Business Economics and Management* 13.3 (2012): 499-517.

Bass, A.N. "From Business Dining To Public Speaking: Tips For Acquiring Professional Presence And Its Role In The Business Curricula." *American Journal of Business Education* 3.2 (2010).

Lockart, M. (2011, June 16). Cubical Etiquette: Sights, Sounds, and Smells. Retrieved from: http://Forbes.com

Heffernan, L. (2013, December 4). Ten Reasons Millennials Need Good Manners. Retrieved from http://grownandflown.com

Schrage, M. "Why Your Looks Will Matter More," *Harvard Business Review* (April 22, 2010).

Hamermesh, Daniel S. "Ugly? You may have a case." *The New York Times Sunday Review* (2011).

Ruetzler, T. et al. "What is Professional Attire Today? A Conjoint Analysis of Personal Presentation Attributes." *International Journal of Hospitality Management* 31.3 (2012): 937-943.

Vitak, J., Crouse, J., and LaRose, R.. "Personal Internet Use at Work: Understanding Cyberslacking." *Computers in Human Behavior* 27.5 (2011): 1751-1759.

Valdez, A.C., Schaar, A.K, and Ziefle, M. "Personality Influences on Etiquette Requirements for Social Media in the Work Context." *Human Factors in Computing and Informatics*. Springer Berlin Heidelberg, 2013. 427-446.

Chapter 5

The IRS: Targeting Americans for Their Political Beliefs. (2013, May 22). Committee on Oversight and Government Reform. http://oversight.house.gov/hearings

Froman, L. "Creating a More Ethical Workplace." *Positive Psychology*. Springer New York, 2013. 161-177.

Meinert, D. "Creating an Ethical Workplace," *HR Magazine* (2014, April 1) Vol. 59 No 4. http://www.eeoc.gov/laws/statutes/

http://www.eeoc.gov/laws/types/

Taylor, B. "Money and the Meaning of Life," *Harvard Business Review* (May 17, 2011).

Etzioni, A. Comparative Analysis of Complex Organizations. (New York: The Free Press, 1961), pp. 4–6.

French, J. R. P., and Raven, B. "The Bases of Social Power." In D. Cartwright (Ed.), *Studies in Social Power*. (Ann Arbor: University of Michigan Press, 1959), pp. 150–167.

Peale, N. V. and Blanchard, K. The Power of Ethical Management. (New York: William Morrow, 1988).

Conner, C. (2013 April 14). "Office Politics: Must You Play? A Handbook for Survival and Success." http://www.forbes.com

Kerby, S. & Burns, C. (2012, July 12). The Top 10 Economics Facts of Diversity in the Workplace. http://www.americanprogress.org

Nishii, L.H., and Susanne, M. "Research Brief: Inside the Workplace: Case Studies of Factors Influencing Engagement of People with Disabilities." (2014).

Chapter 6

"How 'Recession-Proof' Will Millennial Workers Be?" *HR Focus* 86 (March 2009): 6–7.

Leyes, M. "Talkin' 'Bout My Generation," *Advisor Today* 105 (April 2010): 34–38.

Deming, W. E. Quality, Productivity and Competitive Position. (Cambridge, MA: MIT Press, 1982).

Connors, R. and Smith, T. How Did That Happen?: Holding People Accountable for Results the Positive, Principled Way. (New York: Penguin Group, Inc., 2009).

Wakeman, C. Y. "Why Empowerment without Accountability Is Chaos at Work," *Fast Company* (April 7, 2009).

Pugh, K. and Dixon, N., "Don't Just Capture Knowledge—Put It to Work," *Harvard Business Review* (May 2008).

Welch, J. and Welch, S. "Emotional Management," *BusinessWeek* (July 28, 2008).

Staub, R. "Accountability and Its Role in the Workplace," *The Business Journal*. Greensborough, NC January 13, 2005, www.bizjournals.com/triad/stories/2005/01/17/smallb3.html

Chapter 7

Porter, M. E., and Kramer, M.R. "Creating shared value." *Harvard business review* 89.1/2 (2011): 62-77.

Schultz, H. Onward: How Starbucks Fought for its Life without Losing its Soul. (New York: Rodale, Inc., 2011).

Fenn, D. "10 Ways to Get More Sales from Existing Customers," *Inc.* (August 31, 2010), www.inc.com/guides/2010/08/get-more-sales-from-existing-customers.html.

Smith, C. "About Customer Service in a Bad Economy," *The Houston Chronicle* (May 27, 2011), www.smallbusiness.chron.com/customer-service-bad-economy-740.html

Mathews, S. "Why Basic Business Principles Still Matter," September 15, 2010, www.samanthamathews.com/general/why-basic-business-principles-still-matter/

Pleshette, L. A. "Understanding Basic Business Principles," *PowerHomeBiz.com*. Retrieved May 2011, www.powerhomebiz.com/vol50/charan.htm

Sawhney, M., Woldcott, R. C., and Arroniz, I., "The 12 Different Ways for Companies to Innovate." *MIT Sloan Management Review*, Vol. 47, No. 3 (Spring 2006): 75–81.

Von Hippel, E. The Sources of Innovation. (Oxford, UK: Oxford University Press, 1994); and Leonard, D. Wellsprings of Knowledge: Building and Sustaining the Sources of Innovation (Cambridge, MA: Harvard Business School Press, 1998).

Chapter 8

United States Department of Labor, www.dol.gov

Society for Human Resource Management (SHRM), www.shrm.org

Business Research Lab. Houston, Texas, www.busreslab.com

National Labor Relations Board, www.nlrb.gov
http://www.hhs.gov/healthcare/rights/

Wells, S. J. "Layoff Aftermath," HR Magazine (November 2008): 37–41.

Lichtenstein, S. D. and Darrow, J. J. "At-Will Employment: A Right to Blog or a Right to Terminate." Journal of Internet Law, Vol. 11 (March 2008): 1–20.

"Employees Care a Lot More About Performance Reviews Than You May Think," HR Focus 86 (July 2009): 9.

Stanley, T. L. "Union Stewards and Labor Relations," Supervision 71 (February 2010): 3–6.

Wells, S. J. "Getting Paid for Staying Well," HRMagazine 55 (February 2010): 59–62.

Robb, D. "Get the Benefits Message Out," HRMagazine 54 (October 2009): 69–71.

Fogarty, S. "EAPs New Role: A Core Strategic Element," Employee Benefit Advisor 8 (April 2010): 40–46.

Carvin, B. N. "The Great Mentor Match," T&D 63 (January 2009): 46–50.

"Share Baby Boomers' Knowledge With Intergenerational Mentoring," HR Focus 87 (February 2010): 7–13.

Chapter 9

Gallo C. "Why Leadership Means Listening," Businessweek (January 31, 2007), http://www.businessweek.com/smallbiz/content/jan2007/sb20070131_192848.htm

http://articles.economictimes.indiatimes.com/2010-12-10/news/27591286_1_leadership-responsibility-authority. Sangeeth Varghese, Dec 10, 2010, 03.13am IST www.economic times.Com

Dan, R. J. (2010). In the Company of Others: An Introduction to Communication. (New York: Oxford University Press), pp. 157–166.

Susan, Y. "The New Trend in Communication: Silent Listening," Salesopedia. Retrieved May 31, 2011, www.salesopedia.com/index.php/component/content/1863?task=view&Itemid=10479.

Paul, P. "Proxemics in Clinical and Administrative Settings." Journal of Healthcare Management. Vol. 50, No. 3 (May–June 2005): 151–154.

Carter, L. Ideas for adding soft skills education to service learning and capstone courses for computer science students. ACM Technical Symposium on Computer Science Education Proceedings. Dallas, TX. March 9–12, 2011, pp. 517–522.

Chapter 10

Ammann, Philip. "Manners in the Workplace: Time to be Unplugged." (2012).

Benbunan-Fich, R. "The ethics and etiquette of multitasking in the workplace." Technology and Society Magazine, IEEE 31.3 (2012): 15-19.

Pitichat, T. "Smartphones in the workplace: Changing organizational behavior, transforming the future." *LUX: A Journal of Transdisciplinary Writing and Research from Claremont Graduate University* 3.1 (2013): 13.

Ott, A. "How Social Media Has Changed the Workplace," *Fast Company* (November 11, 2010).

Bernstein, C. A., M.D. "Communication: It's Not What It Used to Be Psychiatric News." *American Psychiatric Association*, Vol. 46, No. 6 (March 18, 2011): 3.

Sánchez Abril, P., Levin, A., and Del Riego, A. "Blurred Boundaries: Social Media Privacy and the Twenty-First-Century Employee." *American Business Law Journal* 49.1 (2012): 63-124.

Stricker, B., Dharmasena, M., and Mora, M. "The New Workplace Currency–It's Not Just Salary Anymore: Cisco Study Highlights New Rules for Attracting Young Talent Into the Workplace." *Cisco. com* (2011).

Rafferty, H. R. "Social Media Etiquette: Communicate Behavioral Expectations," *SHRM Online Technology Discipline*, March 24, 2010.

Social Media Acceptable-Use Policy, *HR Magazine* (December 2009).

Study Links Technology to Poor Workplace Manners, *SHRM Online Technology Discipline*, February 2009

Chapter 11

http://www.inc.com/will-yakowicz/leadership-trends-of-2014.html Retrieved May 24, 2014.

Perry, J. E. "Before the Mandate: Cultivating an Organizational Culture of Trust and Integrity." *The American Journal of Bioethics* 13.9 (2013): 42-44.

White, R. and Lippitt, R. Autocracy and Democracy: An Experiential Inquiry (New York: Harper & Brothers, 1960).

Bennis W. and Townsend, R. Reinventing Leadership (New York: William Morrow, 1995).

Carter, M. Z., et al. "Transformational Leadership, Relationship Quality, and Employee Performance During Continuous Incremental Organizational Change." *Journal of Organizational Behavior* 34.7 (2013): 942-958.

Chen, G., et al. "Teams as Innovative Systems: Multilevel Motivational Antecedents of Innovation in R&D Teams." *Journal of Applied Psychology* 98.6 (2013): 1018.

Jablonski, D. *A study of power and individualism in virtual teams: Trends, challenges, and solutions.* Diss. Saybrook Graduate School and Research Center, 2013.

Vanthournout, G., et al. "The Relationship between Workplace Climate, Motivation and Learning Approaches for Knowledge Workers." *Vocations and Learning* (2014): 1-24.

Maslow, A. H. "A Theory of Human Motivation," *Psychological Review*, Vol. 50, No. 4 (July 1943): 370–396.

McClelland, D. The Achieving Society. (New York: Van Nostrand Reinhold, 1961).

Bennis, W. and Nanus, B. Leaders. (New York: Harper & Row, 1985).

Kotter, J. P. "What Leaders Really Do," *Harvard Business Review* 68 (May–June 1990): 103–111.

Berkun, S. Confessions of a Public Speaker. (Sebastopol, CA: O'Reilly Media, Inc., 2009).

Wollan, M. "Strategies for Managers in Handling Workplace Bullying." *Bullying in the Workplace: Symptoms, Causes and Remedies* (2013): 271.

Anderson, C. (2013 June). "How to Give a Killer Presenation," http:///www.hbr.org

Chapter 12

Sugahara, T. and Sugahara, K. "Preventing Workplace Violence and Litigation Through Preemployment Screening and Enforcement of Workplace Conduct Expectations." *Handbook of Forensic Sociology and Psychology* (2014): 185-199.

https://www.osha.gov/SLTC/workplaceviolence/

Katz, N. H., and Flynn, L. T. "Understanding Conflict Management Systems and Strategies in the Workplace: A Pilot Study." *Conflict Resolution Quarterly* 30.4 (2013): 393-410.

De Dreu, C. K. W. "Conflict at Work: Basic Principles and Applied Issues." In S. Zedeck (Ed.), *APA Handbook of Industrial and Organizational Psychology, Vol. 3: Maintaining, Expanding, and Contracting the Organization, APA Handbooks in Psychology.* (Washington, DC: American Psychological Association, 2011), pp. 461–493.

Speakman, J. and Ryals, L. "A Re-evaluation of Conflict Theory for the Management of Multiple, Simultaneous Conflict Episodes." *International Journal of Conflict Management*, Vol. 21, No. 2 (2010): 186–201.

Afzalur R. M. Managing Conflict in Organizations. (New Brunswick, NJ: Transaction Publishers, 2011).

Lovelle B. L. and Lee, R. T. "Impact of Workplace Bullying on Emotional and Physical Well-Being: A Longitudinal Collective Case Study." *Journal of Aggression, Maltreatment & Trauma*, Vol. 20, No. 3 (2011): 344–357.

Chapter 13

Volunteering as a Pathway to Employment Report. (2013 June).

Rothberg, S. (2013 March 28). "80% of Job Openings Are Unadvertised," http://www.CollegeRecruiter.com

Lytle, T. (2013 May 1). Education and Training College Career Centers Create a Vital Link. SHRM. Vol. 58 No.5. http://www.shrm.org

Clay, K. (2014 May 18). "This Service Cleans Up your Social media Profiles for Job Hunting, College Admissions, and Dating." http://payscale.com

Erwin, M. (2013 June 26)."Employers Finding Reasons Not to Hire Candidates on Social Media," http://www.prnewswire.com

Running Background Checks on Job Applicants (May 2014). http://www.nolo.com/employmentlaw/

http://www.onetonline.org/

Golden, J.P. (2014 May 25). Golden Personality Type Profiler. http://www.goldenllc.com

www.myersbriggs.org

UC Santa Cruz Career Center, (2012)."Information interviewing," http://www.careersucsc.edu

The Riley Guide (2014 May). "References & Recommendations," http://www.rileyguide.com

Chapter 14

SHRM Survey Findings: Resumes, Cover Letters and Interviews (2014 April 28). http://www.shrm.org

Bowers, T. (2013 September 16). "10 Mistakes That Could Ruin Your Resume," http://techrepublic.com

"Making Your Résumé E-Friendly: 10 Steps," *The Quick Résumé and Cover Letter Book,* Farr. July 2013, www.careerbuilder.com/Article

"Advanced Résumé Concepts, Electronic Resumes" Kendall, July 2013, www
.reslady.com/electronic.html

"How to Write a Resume Profile," Clark-salaam, March 2014, www.ehow.com

"How to Explain Employment Gaps, Sabbaticals, and Negatives on Your Resume" Vaas, May 2014, www.theladders.com

"Resume Dilemma: Employment Gaps and Job-Hopping, Isaacs, July 2013, www.career-advice.monster.com/resumes-coverletters

Susan Ireland's Resume Site, "How to Explain Unemployment," www
.susanireland.com/resume/how-to-write

Chapter 15

"How to Stand Out from the Crowd and Kick-Start Your Own Recovery," *U.S. News & World Report* 147 (May 2010): 14–16.

National Association of Colleges and Employers, Bethlehem, PA, www.Jobweb
.com

Smith, J. (2013 July 2). "7 Things You Can Do After A Really Bad Job Interview," http://forbes.com

Skorkin, A. The Main Reason Why You Suck at Interviews: Lack of Preparation. *Lifehacker.com*, http://lifehacker.com/5710712/the-main-reason-why-you-suck-at-interviews-lack-of-preparation December 10, 2010.

Chapter 16

Bisharat, J. (2013 September 6). "How to Quit Gracefully: 4 Tips for the Consummate Professional," http://www.huffingtonpost.com

Smith, J. (2012 July 31). "You Quit Your Job. Now They Demand an Exit Interview," http://www.forbes.com

Shea, T. E. "Getting the Last Word," *HR Magazine* 55 (January 2010): 24–25.

Elmer, V. "The Invisible Promotion," *Fortune* (February 7, 2011): 31–32.

www.leanin.org

Bott, N., et al. "Creativity Training Enhances Goal-Directed Attention and Information Processing." *Thinking Skills and Creativity* 13 (2014): 120-128.

Dawson, Patrick, and Constantine Andriopoulos. *Managing change, creativity and innovation.* Sage, 2014.

www.sba.gov

Rae, D. and Woodier-Harris, N.R. "How does Enterprise & Entrepreneurship Education Influence Postgraduate Students' Career Intentions in the New Era Economy?." *Education+ Training* 55.8/9 (2013): 13-13.

Cruikshank, J. "Lifelong learning in the new economy: A great leap backwards." *Canadian Journal of University Continuing Education* 27.2 (2013).

Credits

Index